LOW CARBON, HIGH GROWTH:

LATIN AMERICAN RESPONSES

TO CLIMATE CHANGE

AN OVERVIEW

LOW CARBON, HIGH GROWTH: LATIN AMERICAN RESPONSES TO CLIMATE CHANGE

AN OVERVIEW

Augusto de la Torre
Pablo Fajnzylber
John Nash

THE WORLD BANK
Washington, D.C.

©2009 The International Bank for Reconstruction and Development / The World Bank
1818 H Street, NW
Washington, DC 20433
Telephone: 202-473-1000
Internet: www.worldbank.org
E-mail: feedback@worldbank.org

1 2 3 4 12 11 10 09

This volume is a product of the staff of the International Bank for Reconstruction and Development / The World Bank. The findings, interpretations, and conclusions expressed in this volume do not necessarily reflect the views of the Executive Directors of The World Bank or the governments they represent.

The World Bank does not guarantee the accuracy of the data included in this work. The boundaries, colors, denominations, and other information shown on any map in this work do not imply any judgement on the part of The World Bank concerning the legal status of any territory or the endorsement or acceptance of such boundaries.

Rights and Permissions
The material in this publication is copyrighted. Copying and/or transmitting portions or all of this work without permission may be a violation of applicable law. The International Bank for Reconstruction and Development / The World Bank encourages dissemination of its work and will normally grant permission to reproduce portions of the work promptly.

For permission to photocopy or reprint any part of this work, please send a request with complete information to the Copyright Clearance Center, Inc., 222 Rosewood Drive, Danvers, MA 01923, USA; telephone: 978-750-8400; fax: 978-750-4470; Internet: www.copyright.com.

All other queries on rights and licenses, including subsidiary rights, should be addressed to the Office of the Publisher, The World Bank, 1818 H Street, NW, Washington, DC 20433, USA; fax: 202-522-2422; e-mail: pubrights@worldbank.org.

ISBN: 978-0-8213-7619-5
eISBN: 978-0-8213-7921-9
DOI: 10.1596/978-0-8213-7619-5

Library of Congress Cataloging-in-Publication Data has been requested.

Cover design: Naylor Design.

Contents

Acknowledgments vii

Preface ix

1. Introduction 1

2. Climate Change Impacts in Latin America and the Caribbean 1

3. The Need for a Coordinated, Effective, Efficient, and Equitable Global Response 15

4. LAC's Potential Contribution to Global Mitigation Efforts 20

5. Policies for a High-Growth, Low-Carbon Future 40

6. Summary and Conclusions 59

Annex 1: Mitigation Potential by Country and Type of Emissions 61

Annex 2: Potential Annual Economic Impacts of Climate Change in CARICOM Countries
circa 2080 (in millions 2007 US$) 65

Bibliography 66

Endnotes 72

Acknowledgments

LOW CARBON, HIGH GROWTH: Latin American Responses to Climate Change is the product of a collaborative effort of two units of the Latin American and the Caribbean Region of the World Bank: the Office of the Chief Economist and the Sustainable Development Department. The report was prepared by a core team led by Pablo Fajnzylber and John Nash, and comprising Veronica Alaimo, Javier Baez, Svetlana Edmeades, Christiana Figueres, Todd Johnson, Irina I. Klytchnikova, Andrew Mason, and Walter Vergara. Ana F. Ramirez and Carlos Felipe Prada Pombo provided valuable research assistance to the team.

The team greatly benefited from background papers and other inputs prepared for this report by the following individuals: Veronica Alaimo, Carlos E. Arce, Juliano J. Assunçao, Javier Baez, Brian Blankespoor, Eduardo Bitran Colodro, Benoit Bosquet, Flavia Chein Feres, Shun Chonabayashi, Alejandro Deeb, Uwe Deichmann, Ariel Dinar, Manuel Dussan, Vladimir Gil, Harry de Gorter, Hilda R. Guerrero Rojas, David R. Just, Erika Kliauga, Donald F. Larson, Humberto Lopez, Carla della Maggiora, Andrew Mason, Robert Mendelsohn, Bekele Debele Negewo, Carmen Notaro, Paul Procee, Claudio Raddatz, Pedro Rivera, Pasquale L. Scandizzo, Sebastian Scholz, Shaikh Mahfuzur Rahman, Yacov Tsur, Dominique Van Der Mensbrugghe, Denis Medvedev, Felix Vardy, Antonio Yunez Naude, Steven Zanhiser, Natsuko Toba, Adriana Valencia, and Seraphine Haeussling.

This Overview (Volume I) of the report was prepared by Augusto de la Torre, Pablo Fajnzylber, and John Nash. The authors of the chapters in the forthcoming Volume II are as follows: chapter 1, Fajnzylber and Nash; chapter 2, Nash and Vergara; chapter 3, Nash, Edmeades, Baez, and Mason; chapter 4, Fajnzylber and Figueres; chapter 5, Fajnzylber and Alaimo; chapter 6, Johnson and Klytchnikova.

Special thanks go to Laura Tuck for her careful reading of drafts of the documents and for her insightful comments and suggestions on both substantive and editorial levels. Excellent guidance and advice was also received from peer reviewers Marianne Fay and Charles Feinstein, as well as Makhtar Diop, Mac Callaway, Jocelyne Albert, and Carlos Nobre. And last, but certainly not least, we would like to thank Susan Goldmark for proposing the idea of doing this regional report on climate change.

Preface

A GLOBAL FINANCIAL AND ECO-
NOMIC CRISIS of unprecedented
dimensions was unfolding at the time
of this writing. The urgency, immedi-
acy, and staggering magnitude of the
challenges posed by such a crisis have the potential to
crowd out efforts aimed at addressing the challenges
of global warming that are discussed in detail in this
report. The capacity of political leaders and of national
and supranational institutions to deal with major
global threats is, after all, not unlimited. It would be,
therefore, naïve to think that the world's ability to
tackle simultaneously the breakdown of financial markets and the threats posed by global warming is free of
tensions and trade-offs. These two global menaces are
of such far-reaching implications for mankind, however, that it would be imprudent to allow the shorter-
term emergency of the global financial crisis and
economic downturn to unduly deflect policy attention
away from the longer-term dangers of climate change.
The challenge clearly is to find common ground and
to identify and pursue as many policies as feasible that
can deliver progress on both fronts simultaneously.
This is possible in principle, but not easy to achieve in
practice.

In effect, the world economic slump will be associated with a fall in private investment, including climate-friendly investment. The latter may tend to
suffer disproportionately in the current context, given
that the price of fossil fuels has fallen dramatically relative to alternative, clean sources of energy. Not surprisingly, utilities already seem to be making
significant reductions in their investments in alternative energy, and there is already a reduction in the
flow of project finance devoted to low-carbon energy
projects. The expectation that a low relative price of
fossil fuels is here to stay might not only deter investment in low-carbon technology, it could also induce
substitution in consumption in favor of cheaper but
dirtier energy. For example, low gasoline prices could
deflate the momentum toward hybrid vehicles, particularly in North America. With lower economic
growth worldwide, furthermore, greenhouse gas
(GHG) emissions could experience a *cyclical* decline;
this might create political incentives to postpone policy efforts to bring down the emissions *trend*. In all,
the global financial and economic crisis could lead to a
shortening of policy horizons that might induce a
shift toward a more carbon-intensive growth path.
This shift would only increase the difficulty and raise
the costs of reducing GHG emissions down the line.

Experience with previous financial crises in emerging economies suggests that tradeoffs often arise
between long-term environmental concerns and
short-term macroeconomic policy responses.[1] In particular, as competing claims rise on shrinking bud-
getary resources during a crisis, budget cuts tend to
affect to a larger extent the provision of public services

that are considered to be a "luxury"—that is, services whose immediate impact on the people or sectors affected by the emergency is perceived to be low and only indirect. In developing countries, these often include such items as forest conservation or the protection of ecosystems. According to an IMF paper,[2] for example, in the aftermath of the Asian and Russian crises, Brazil reduced public expenditures (excluding wages, social security benefits, and interest payments) for 1999 by 11 percent in nominal terms with respect to 1998. However, some key Amazon environmental programs were reduced by much more than the average. The Brazilian Institute for the Environment and Natural Renewable Resources (IBAMA), for instance, experienced a budget cut of 71 percent with respect to originally approved funding, and of 46 percent compared to 1998. There are also indications that this phenomenon went beyond the federal level. Brazilian states and municipalities, faced with the need to produce "primary surpluses," were not able to compensate for the cuts in federally funded environmental programs in the Amazon.[3]

If leaders at the national and international levels are visionary, they can avoid falling into the trap of sacrificing environmental sustainability to short-term macroeconomic necessities, and can take advantage of opportunities to address climate change concerns. In particular, policies and programs to address today's pressing problems can be designed and implemented with a long-term horizon. Sometimes, these decisions can be win-win. But sometimes, there will be trade-offs. For example, private investment in, and consumption of, clean energy will be stimulated by a relative increase in the price of fossil fuels; this can be encouraged through a combination of regulations, taxes, carbon-trading schemes, and subsidies. But making firms pay to pollute and forcing households to consume more expensive, if cleaner, energy are not popular in times of economic recession. Tilting private-sector activity in a sustainable fashion toward low-carbon choices thus calls for carefully managed politi-

cal compromises and sound judgment on the part of policy makers to ensure that long-term considerations are not neglected for political expediency.

Greater scope for synergies is likely to be found in the area of public investment. Massive public investment programs will have to be part of the fiscal stimulus required to deal with the global economic crisis, especially in developed countries and high-saving emerging economies. Appropriately designed and implemented, these programs can generate win-win dynamics and outcomes, simultaneously advancing the causes of supporting economic recovery while helping to encourage growth in areas that minimize or mitigate the impact on climate change. Moreover, countries that manage to effect the transition from a high-carbon to a low-carbon economy during the economic slump can enjoy "first-mover advantages," that is, a greater competitive ability to promote long-term growth beyond the cyclical downturn. As a result, the current financial crisis can actually create a unique opportunity for a new deal for the 21st century, focused on low-carbon growth. The declared vision for environmental sustainability and energy security of the recently elected government in the United States adds hope in this regard. A "green recovery"—that is, a virtuous interaction among job creation, growth resumption, and low-carbon-oriented public investments and policy actions—is a worthy option and arguably the only sensible option for the world community at this juncture. Such an option can be turned into reality if leaders and political systems rise to the occasion.

Laura Tuck
Director, Sustainable Development Department
Latin America and the Caribbean Region
The World Bank

Augusto de la Torre
Chief Economist
Latin America and the Caribbean Region
The World Bank

Low Carbon, High Growth: Latin American Responses to Climate Change

1. Introduction

Based on analysis of recent data on the evolution of global temperatures, snow and ice covers, and sea level rise, the Intergovernmental Panel on Climate Change (IPCC) has recently declared that "warming of the climate system is unequivocal."[4] Global surface temperatures, in particular, have increased during the past 50 years at twice the speed observed during the first half of the 20th century.

The IPCC has also concluded that with 95 percent certainty the main drivers of the observed changes in the global climate have been anthropogenic increases in greenhouse gases (GHG).[5] Models of the evolution in global temperatures that take into account the effects of man-made emissions of greenhouse gases (the pink paths in map 1) match much better with actual recorded temperatures (the black lines) than do models that do not incorporate these effects.[6] The conclusion is inescapable that, as man-made emissions have accumulated in the atmosphere, they have caused temperatures to increase.

While the greenhouse effect is a natural process without which the planet would probably be too cold to support life, most of the increase in the overall concentration of GHGs observed since the Industrial Revolution has been the result of human activities, namely the burning of fossil fuels, changes in land use (conversion of forests into agricultural land), and agriculture (the use of nitrogen fertilizers and livestock-related methane emissions).[7]

Looking forward, the IPCC predicts that global GHG emissions will increase by as much as 90 percent between 2000 and 2030 if no additional climate change mitigation policies are implemented. As a result, under "business as usual" scenarios, global temperatures could increase by as much as 1.7°C by 2050 and by up to 4.0°C by 2100. Actual emissions during recent years, however, have matched or exceeded the IPCC's most pessimistic forecasts (figure 1). Taking this into account, Stern (2008) predicts that the stock of GHG in the earth's atmosphere could increase from the current level of 430 parts per million to 750 by 2100.[8] This would imply that global warming with respect to preindustrial times would exceed 4°C with an 82 percent probability and would rise above 5°C with a 47 percent probability.

2. Climate Change Impacts in Latin America and the Caribbean

The "unequivocal" warming of the climate system reported by the IPCC is already affecting Latin America's climate. Temperatures in Latin America increased by about 1°C during the 20th century, while sea-level rise has reached 2–3 mm/yr since the 1980s. Changes in precipitation patterns have also been observed, with some areas receiving more rainfall (southern Brazil, Paraguay, Uruguay, northeast Argentina, and northwest Peru), and others less (southern Chile, southwest Argentina, and southern Peru). Finally, extreme weather events have become more common in

MAP 1

World Actual and Modeled Average Temperatures, by Region, 1900–2000

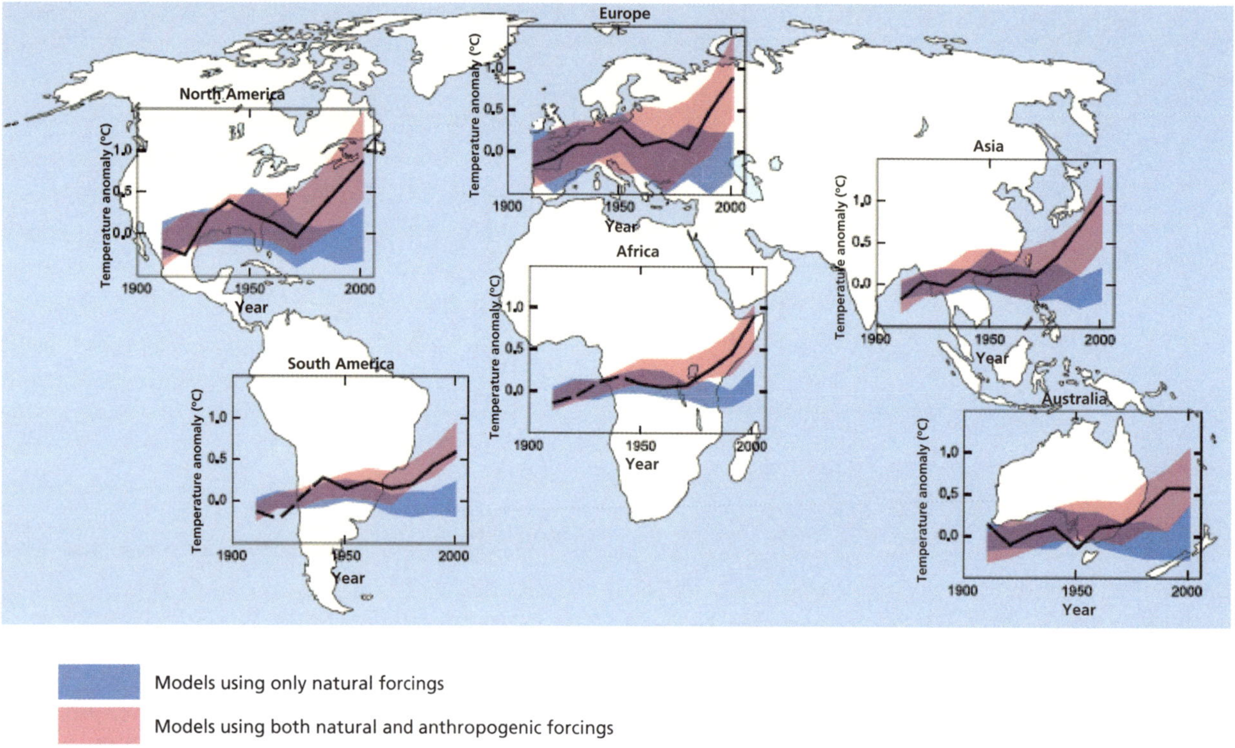

Models using only natural forcings

Models using both natural and anthropogenic forcings

Source: Climate Change 2007: Synthesis Report. Contribution of Working Groups I, II and III to the Fourth Assessment Report of the Intergovernmental Panel on Climate Change. Figure SPM.4. IPCC, Geneva, Switzerland.

FIGURE 1

Observed Global CO$_2$ Emissions Compared with Emissions Scenarios and Stabilization Trajectories

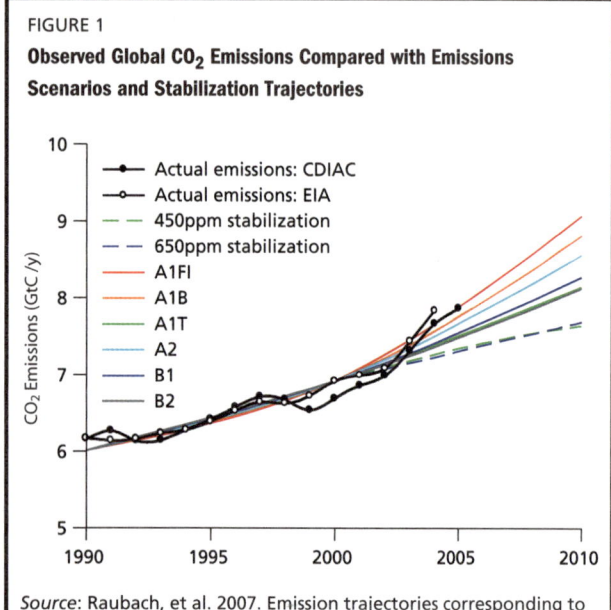

Source: Raubach, et al. 2007. Emission trajectories corresponding to the main scenarios studied by the IPCC's Special Report on Emission Scenarios (2001).
Note: The curves shown for scenarios are averages over available individual scenarios in each of the six scenario families, and differ slightly from "marker" scenarios. Further details on each scenario and sources of data are in the attached endnote.[9]

several parts of the region, including more periods of intense rainfall and consecutive dry days.[10]

Ecosystems are already suffering negative effects from ongoing climate change in LAC

Apart from some possible positive effects on crop yields in the Southern Cone, the impacts so far have been profoundly negative, already affecting some of the unique features and ecosystems of the region. Based on their irreversibility, their importance to the ecosystem, and their economic cost, four impacts stand out as being of special concern. These Climate Ecosystem Hotspots are (a) the warming and eventual disabling of mountain ecosystems in the Andes; (b) the bleaching of coral reefs leading to an anticipated total collapse of the coral biome in the Caribbean basin; (c) the damage to vast stretches of wetlands and associated coastal systems in the Gulf of Mexico; and (d) the risk of forest dieback in the Amazon basin. In this section of the report, we initially present evidence on the first three of these processes, which are ongoing, as well as on the increasing damage from tropical storms, another current phenomenon. We then address future expected climate trends and their possible impacts, including the above-mentioned risk of Amazon dieback, as well as other impacts on natural and human systems.

The *melting of the Andean glaciers with damage to associated ecosystems* has been going on for some years, driven by the higher rates of warming that have been observed at higher altitudes (figure 2).[11] An analysis of trends in temperature (Ruiz-Carrascal et al. 2008) indicates possible increases on the order of 0.6°C per decade, affecting the northern, more humid section of the Andes. Many of the smaller glaciers (less than 1

square kilometer in area) have declined in surface area. For example, Bolivia's Chacaltaya Glacier has lost most (82 percent) of its surface area since 1982 (Francou et al. 2003). High mountain ecosystems, including unique high altitude wetlands ("*paramos*") associated with the glaciers, are among the environments most sensitive to climate change. These ecosystems provide numerous and valuable environmental goods and services, and drastic reductions in populations of mountain flora and fauna have already been observed in recent years.

Another serious environmental impact already observable is the *bleaching of coral reefs* in the Caribbean. Coral reefs are home to more than 25 percent of all marine species, making them the most bio-

FIGURE 2
Retreat of the Chacaltaya Glacier in Bolivia

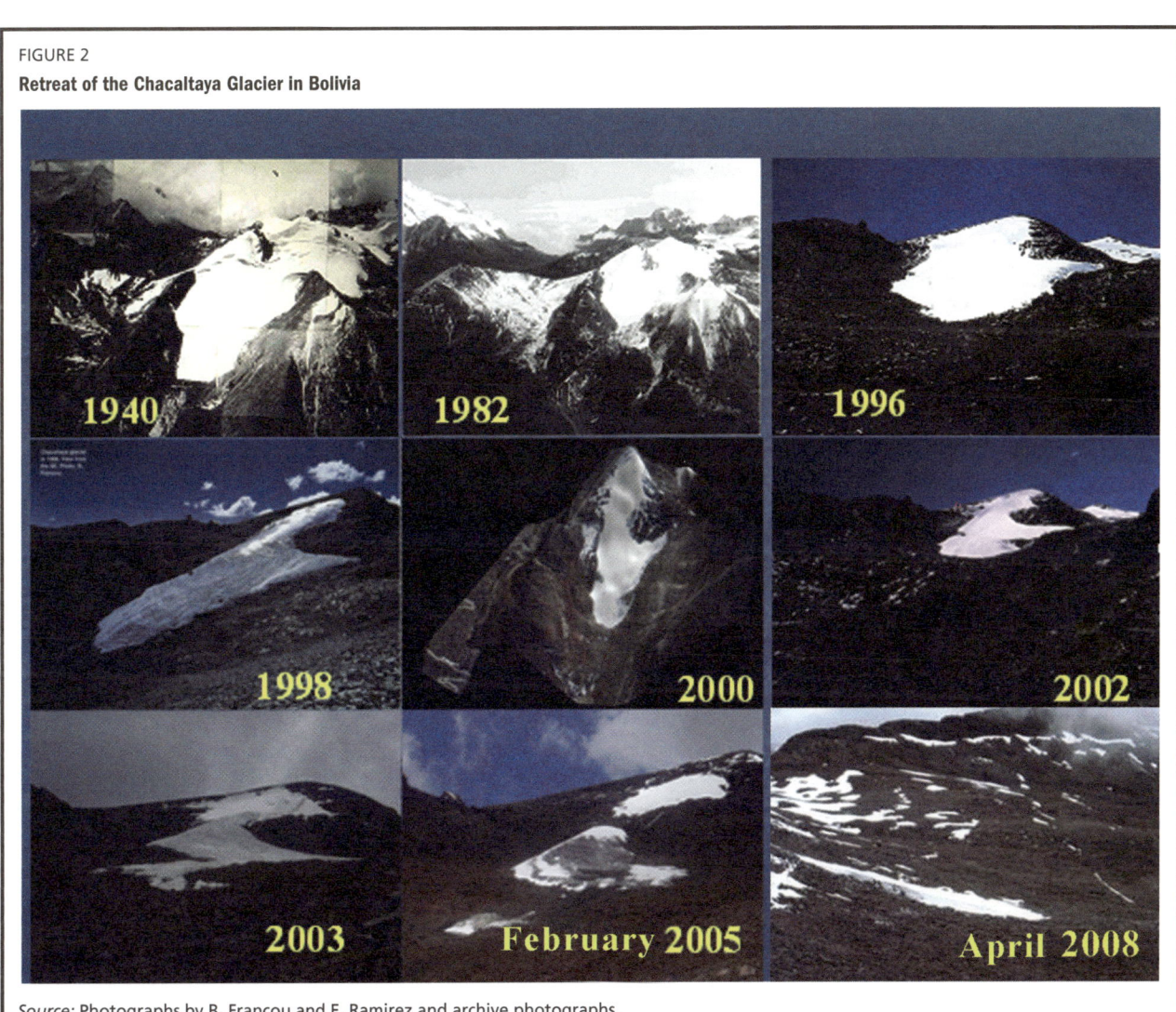

Source: Photographs by B. Francou and E. Ramirez and archive photographs.

logically diverse of marine ecosystems, and an analog to rainforests on land ecosystems. In the case of the Caribbean, coral reefs are hosts to fish nurseries for an estimated 65 percent of all species in the region, so their survival is critical to the ecology of the ocean in this region. When stressed by heat, corals expel the microscopic algae living symbiotically in their tissues. If this is a one-time event, it is not necessarily fatal, but repeated episodes will kill the reef. Consistent increases in sea surface temperatures have led to several recent bleaching events (1993, 1998, 2005), the latest of which caused widespread bleaching throughout the region.

Damage to the Gulf Coast wetlands in Mexico is yet another serious ongoing concern. Global circulation models agree that the Gulf of Mexico is the most vulnerable coastal area in the region for impacts from climate change, and Mexico's three national communications (NCs) to the UNFCCC[12] have documented ongoing damage, raising urgent concerns about their integrity. Wetlands in this region are currently suffering from anthropogenic impacts derived from land use changes, mangrove deforestation, pollution, and water diversion. These make the ecosystem even more vulnerable to climate change impacts, including the reduction in rainfall of up to 40 percent that is forecast by 2100 (P. C. D. Milly et al. 2005). Total mangrove surface is disappearing at a rate of 1–2.5 percent per year. Wetlands provide many environmental services, including the regulation of hydrological regimes, protection of human settlement from floods and storms, sustenance for many communities settled along the coast, and habitats for waterfowl and wildlife. These wetlands possess the most productive ecosystem in that country and one of the richest on earth.[13] About 45 percent of Mexico's shrimp production, for example, originates in the Gulf wetlands, as do 90 percent of the country's oysters and no less than 40 percent of commercial fishing volume. While other coastal areas in the LAC region will also be prone to similar impacts, the biological and economic value of the Gulf wetlands justifies their identification as a particularly important climate hotspot.

Data are also suggestive of a trend underway of *more and/or stronger storms and weather-related natural disasters in the region*. Estimates of the macroeconomic cost of climatic natural disasters suggest that on average, each of them causes a 0.6 percent reduction in real GDP per capita. To the extent that, since the 1990s, such events have taken place on average once every three years—compared to once every four years in the period since 1950—their average impact on the affected countries would be a 2 percent reduction in GDP per capita per decade (Raddatz 2008).[14]

Latin Americans are well aware of the high toll taken by extreme weather events. In 1999, for example, 45,000 people were killed in floods and mudslides in República Bolivariana de Venezuela, while Hurricane Mitch in 1998 killed at least 11,000 and perhaps 19,000 across Central America and Mexico. One report calculated the economic damage in Honduras at US$3.8 billion—two-thirds of GDP. More recently, Hurricane Wilma in 2005, the strongest Atlantic hurricane on record, damaged 98 percent of infrastructure along the southern coast of Mexico's Yucatan Peninsula, home to Cancun, and inflicted an estimated US$1.5 billion loss on the tourism industry.

Recent reviews of hurricane activity over time (Hoyos et al. 2006; Webster and Curry 2006) point to trends in the intensification of hurricanes. Of particular significance is the recent increase in Mesoamerican landfalls since 1995 after an extended quiet regime of nearly 40 years. In 2004, for the first time ever, a hurricane formed in the South Atlantic and hit Brazil. And the year 2005 saw the number of hurricanes in the North Atlantic hit 14, a historic high. Four of the ten most active years for hurricane landfalls have occurred in the last 10 years, and 2008 saw Cuba, Haiti, and other islands devastated by multiple hits. This raises the question of whether we are already seeing an impact of climate change that will increase the expected damages in the region. In fact, following Hurricane Katrina, U.S. risk-modeling companies raised their estimation of the probability of a similar event from once every 40 years to once every 20 years as a result of the warming of water temperatures in the North Atlantic Basin. Taking all kinds of climate-

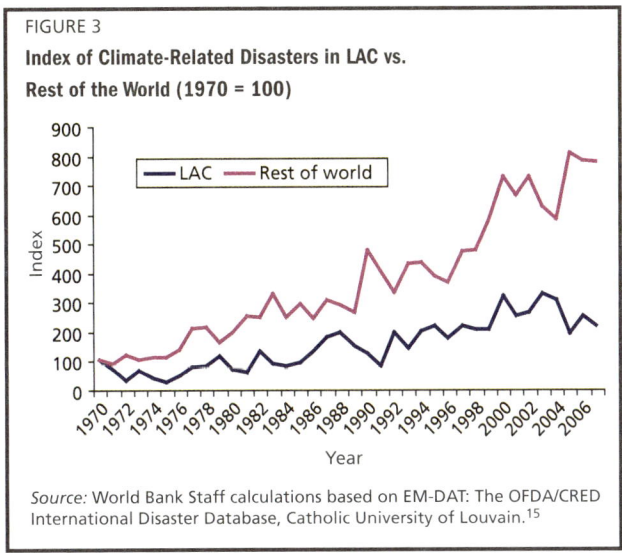

FIGURE 3

Index of Climate-Related Disasters in LAC vs. Rest of the World (1970 = 100)

Source: World Bank Staff calculations based on EM-DAT: The OFDA/CRED International Disaster Database, Catholic University of Louvain.[15]

related disasters together, there appears to be a positive trend over the last few decades, although less marked in LAC than in the rest of the world (figure 3).

As climate change intensifies, more serious consequences are likely in the future

The IPCC's Fourth Assessment Report predicts that under business-as-usual scenarios, temperature increases in LAC with respect to a baseline period of 1961-1990 could range from 0.4°C to 1.8°C by 2020 and from 1°C to 4°C by 2050 (Magrin et al. 2007). In most of the region, the expected annual mean warming is likely to be higher than the global mean, the exception being the southern part of South America (Christensen et al. 2007). These projections, derived from global circulation models, also forecast changing precipitation patterns across the region, although in many subregions there is much less agreement among the models on the direction and magnitude of changes in rainfall than on the change in temperature. In Central America, for example, while most models do predict lower mean precipitation in all seasons, there is a possibility that this could be compensated by increased rainfall during hurricanes, which is not well captured in most general circulation models.[16]

Notwithstanding the high uncertainty regarding future rainfall patterns in some areas, there are strong indications that climate change may magnify extremes already observed across the region. Thus, as illustrated in the first four panels of map 2 (see p. 6), it appears that many areas with a current high exposure to droughts or flood risks would in the future have to deal with respectively even drier conditions and more intense rainfall.

In particular, this would be the case in all the high-drought-risk areas of Chile, Mexico, Guatemala, and El Salvador, for which the predictions of at least seven to eight global climate models indicate that by 2030 the number of consecutive dry days will increase and heat waves will become longer. Similarly, between 47 and 100 percent of the high-flood-risk areas of Argentina, Peru, and Uruguay are expected to become even more exposed to intense rainfall. True, there are still considerable differences in the specific regional projections derived from various global climate models. However, as illustrated in the four panels of map 2 showing concordance (see p. 7), for most of the examples above, the majority of the available climate models coincide at least in the sign of their predictions.

Climate change will also lead to a rising sea level, which will affect all coastal areas. Sea level is forecast by the Fourth Assessment of the IPCC (2007) to rise by 18 to 59 centimeters in the current century from thermal expansion as the air warms from glacial melt (mainly in Greenland and Antarctica) and from changes in territorial storage capacity. There remains, however, considerable scientific uncertainty over the state of the Greenland Ice Sheet, which holds water sufficient to raise sea level by 7 meters, and the Antarctic, which could raise sea level by 61 meters if fully melted. Small changes in volume of these could have a significant impact. So, while large-scale rise in sea level is not highly likely in periods less than centuries, there remains much uncertainty, and recent evidence does point to more rapid increases than in the IPCC's Third Assessment Report (Dasgupta et al. 2007).

Damages to ecosystems will be even more serious in the future...

The impacts in the future on ecosystems and human society of such changes could be profound. Perhaps the most disastrous impact, if it occurs, will be a *dramatic*

MAP 2

Expected Climate Risks and Measures of Model Concordance in LAC, 2030

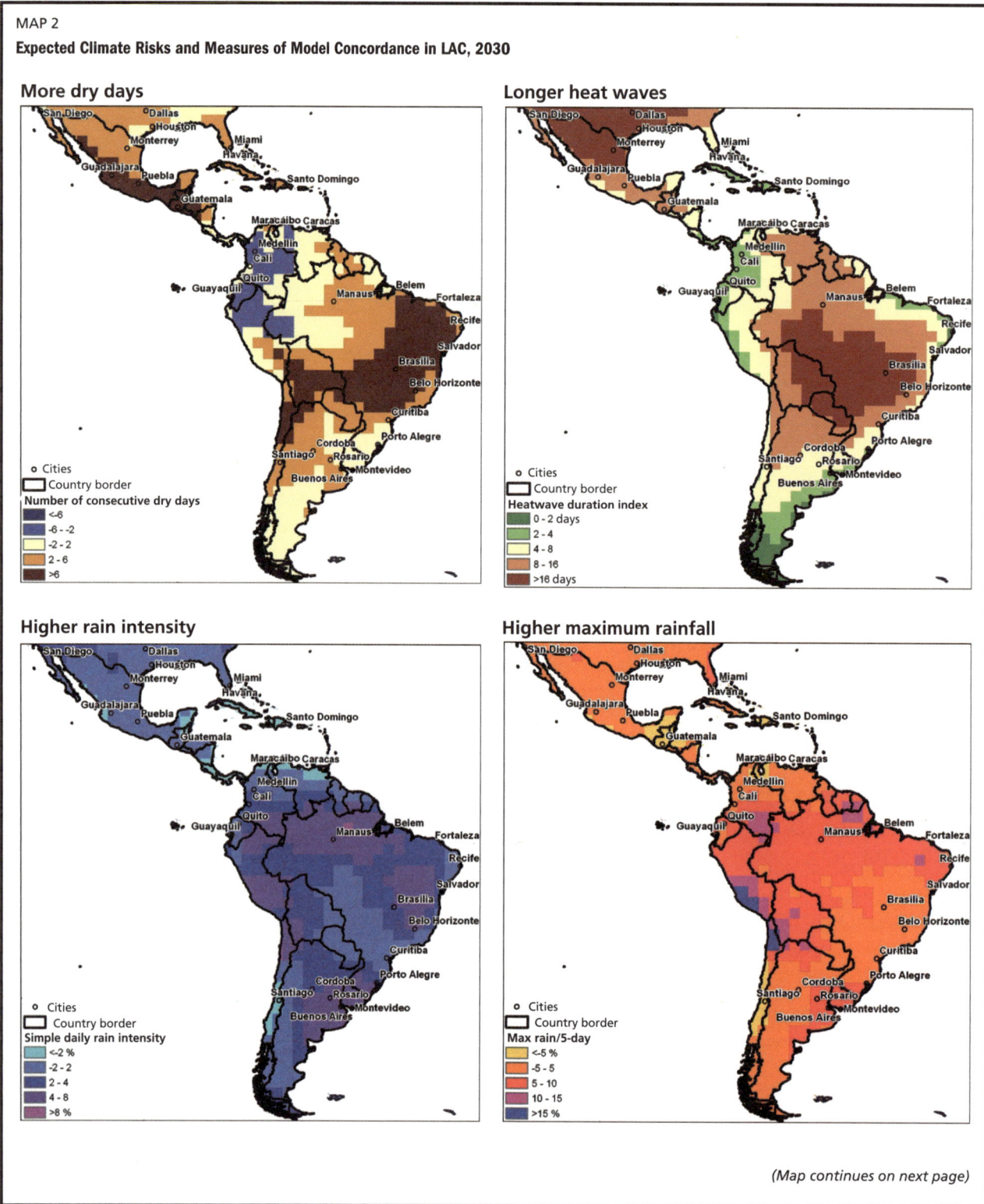

(Map continues on next page)

MAP 2
(continued)

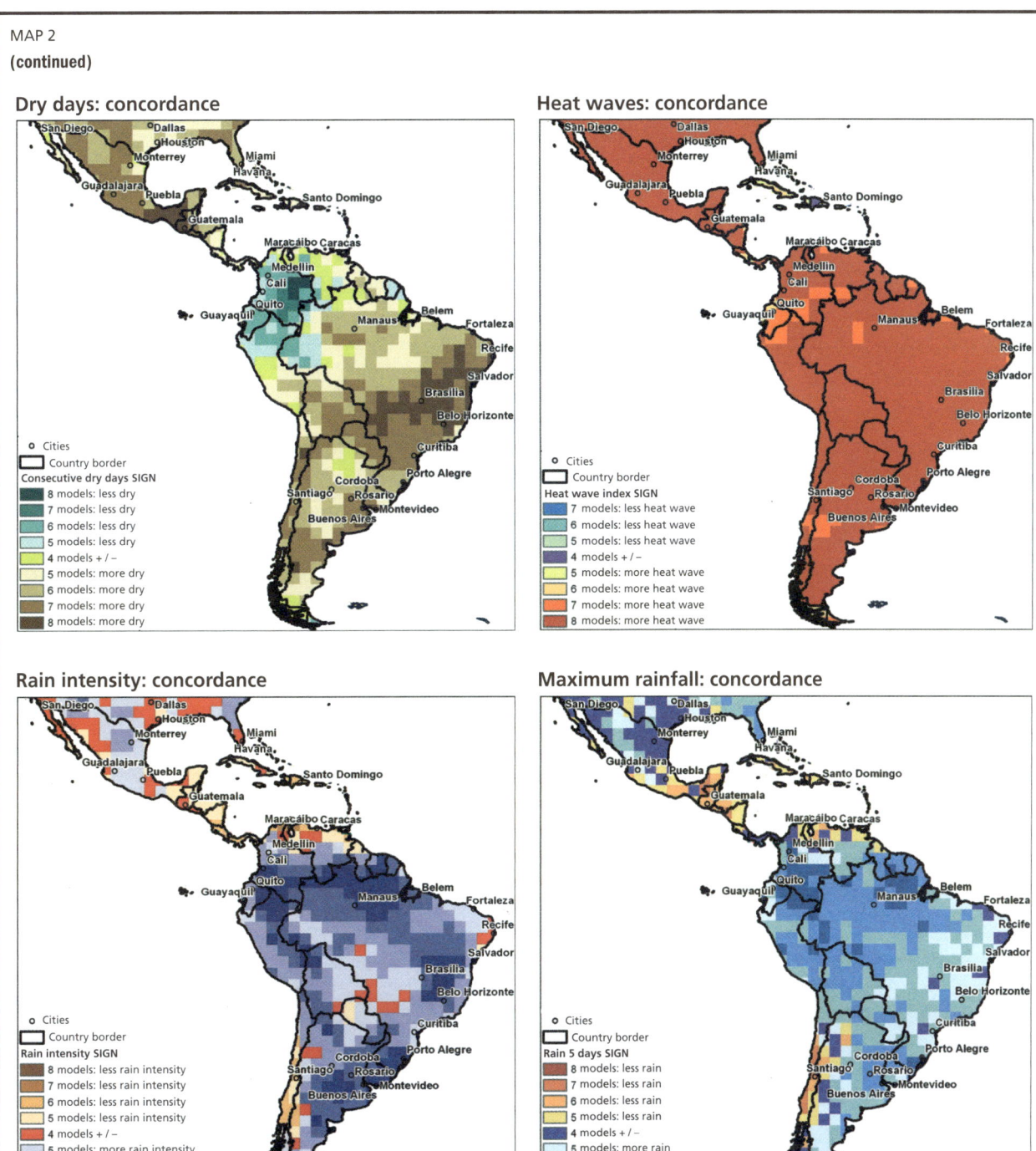

Dry days: concordance

Heat waves: concordance

Rain intensity: concordance

Maximum rainfall: concordance

Source: World Bank Staff calculations using eight global circulation models. Lower four maps indicate concordance (agreement) among forecasts of different models. Model concordance is measured by the number of models whose predictions for changes in temperatures or rainfall are of the same sign.

dieback of the Amazon rainforest, with large areas converted to savannah. Most Dynamic Global Vegetation Models (DGVM) based on the IPCC emission scenarios show a significant risk of climate-induced forest dieback toward the end of the 21st century in tropical, boreal, and mountain areas, and some General Circulation Models predict a drastic reduction in rainfall in the western Amazon.[17] While there is as yet no consensus in the scientific community regarding the likelihood and extent of the possible dieback of the Amazon, the Technical Summary of the Fourth Assessment Report of the IPCC indicates a potential Amazon loss of between 20 and 80 percent as a result of climate impacts induced by a temperature increase in the basin of between 2.0 and 3.0°C. The credibility of this kind of scenario was reinforced in 2005, when large sections of southwestern Amazonia experienced one of the most intense droughts of the last 100 years. The drought severely affected human populations along the main channel of the Amazon River and its western and southwestern tributaries.

The Amazonian rainforest plays a crucial role in the climate system. It helps to drive atmospheric circulation in the tropics by absorbing energy and recycling about half of the rainfall that falls upon it. Furthermore, the region is estimated to contain about 10 percent of the global stock of carbon stored in land ecosystems, and to account for 10 percent of global net primary productivity (Melillo et al. 1993).[18] Moisture injected by the Amazon ecosystem into the atmosphere also plays a critical role in the precipitation patterns in the region. Disruptions in the volumes of moisture coming from the Amazon basin could trigger a process of desertification over vast areas of Latin America and even in North America (Avissar and Werth 2005). The IPCC also indicates a likelihood of major biodiversity extinctions as a consequence of Amazon dieback.

Even apart from the huge loss of biodiversity from such cataclysmic changes as Amazon dieback, climate change will *threaten the rich biodiversity of the LAC Region more generally.* Of the world's 10 most biodiverse countries, 5 are in LAC: Brazil, Colombia, Ecuador, Mexico, and Peru, and this list also comprises 5 of the 15 countries whose fauna are most threatened with extinction.[19] The single most biologically diverse area in the world is the eastern Andes. Around 27 percent of the world's mammals live in LAC, as do 34 percent of its plants, 37 percent of its reptiles, 43 percent of its birds, and 47 percent of its amphibians. Forty percent of the plant life in the Caribbean is unique to this area. Climate change is likely to drastically affect the survival of species, as breeding times and distributions of some species shift.[20] Arid regions of Argentina, Bolivia, and Chile, along with Mexico and central Brazil, are likely to experience severe species loss by 2050 using mid-range climate forecasts (Thomas and others 2004). Mexico, for example, could lose 8–26 percent of its mammal species, 5–8 percent of its birds, and 7–19 percent of its butterflies. Species living in cloud forests will become vulnerable, as the warming causes the cloud base to rise in altitude. In the cloud forest of Monteverde in Costa Rica, this kind of change is already being observed, as reductions in the number of mist days have been associated with decrease in populations of amphibians, and probably also birds and reptiles (Pounds et al. 1999). Amphibians are especially susceptible to climate change. Species that are both threatened (according to the Red List of the IUCN) and climate change-susceptible inhabit areas of Mesoamerica, northwestern South America, various Caribbean Islands, and southeastern Brazil (map 3). Among birds, the families that are highly susceptible and are endemic to Latin America are *Turdidae* (thrushes, 60 percent of which are classified as highly susceptible), *Thamnophilidae* (antbirds, 69 percent highly susceptible), *Scolopacidae* (sandpipers and allies, 70 percent highly susceptible), *Formicariidae* (ant thrushes and ant pittas, 78 percent highly susceptible) and *Pipridae* (manakins, 81 percent highly susceptible).[21]

...and socioeconomic damages will be high as well

Climate change is likely to also cause severe negative impacts on socioeconomic systems. Some of these socioeconomic impacts will be due to the direct effects of climate on human activities, while others will be intermediated through the impact that the climate will have on ecosystems which provide economically significant services. Among the economic sectors, the

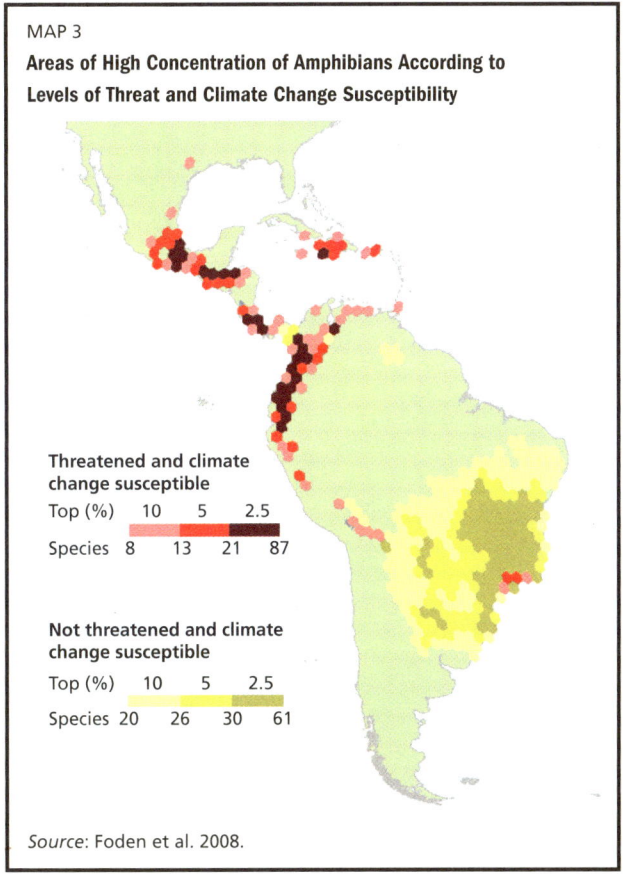

MAP 3

Areas of High Concentration of Amphibians According to Levels of Threat and Climate Change Susceptibility

Threatened and climate change susceptible

Top (%)	10	5	2.5
Species	8	13	21	87

Not threatened and climate change susceptible

Top (%)	10	5	2.5
Species	20	26	30	61

Source: Foden et al. 2008.

between climate and farm production is quantified, forecasts of future climatic changes (in temperatures and precipitation) can be used to predict how farmers will respond. Endogenous choices by farmers to own livestock, choose crop types, pick livestock species, determine herd size, and install irrigation can all be examined with these data. The standing hypothesis is that these choices are sensitive to climate. The models also examine how land values—as a measure of overall profitability—vary with climate. Applications of this so-called Ricardian approach to data from Mexico and seven South American countries reveal that indeed, land values are sensitive to climate and tend to fall with higher temperatures and higher precipitation, over ranges of these variables that are relevant to Latin America. These studies also find—somewhat contrary to expectations—that in percentage terms, small farms are not more severely impacted than large, perhaps because the larger farms tend to be more specialized in temperate (heat-intolerant) crops and livestock, and therefore less adaptable.[22] Of course, small farmers living close to the margin of subsistence will suffer greater hardship than will larger farmers from a similar percentage decline in production.

In the case of the South American farms studied in this report, average simulated revenue losses from climate change in 2100 are estimated to range from 12 percent for a mild climate change scenario to 50 percent in a more severe scenario, even after farmers undertake adaptive reactions to minimize the damage.[23] (Of course, these kinds of studies cannot take into account potential adaptive responses using future technological developments.) Another study applying similar techniques to Mexico forecasts that that country would be heavily impacted, with a virtually total loss of productivity for 30–85 percent of all farms, depending on the severity of warming.[24] Yet it is worth noting that across countries and even within the same country, the impacts are likely to vary substantially from one region to the next. (Map 4 reports the results for small farms, which have a pattern of impacts similar to that for large farms.) Even in hard-hit Mexico, some regions are forecast to benefit. Across the continent of South America, losses are generally forecast to be higher nearer the equator, with

one likely to suffer the most direct and largest impact from gradual changes in temperature and precipitation is agriculture. Also important, at least from a local perspective, are the economic and social impacts of the expected increase in the frequency and/or intensity of hurricanes and tropical storms, the disappearance of tropical glaciers in the Andes, the increase in the rate of sea-level rise, the bleaching and eventual dieback of coral reefs in the Caribbean, possible water shortages created by changes in rainfall patterns, and the expected increase in mortality and morbidity rates derived from climate-related changes in the prevalence of various diseases.

Agricultural productivity could suffer a precipitous fall in many regions. One of the leading approaches to estimating the long-run impacts of climate change on agriculture takes advantage of individual data on large cross-sections of farmers. By matching farms to climates, and adjusting for other characteristics, one can examine how climate influences farm decisions and economic returns to farming. Once the relation

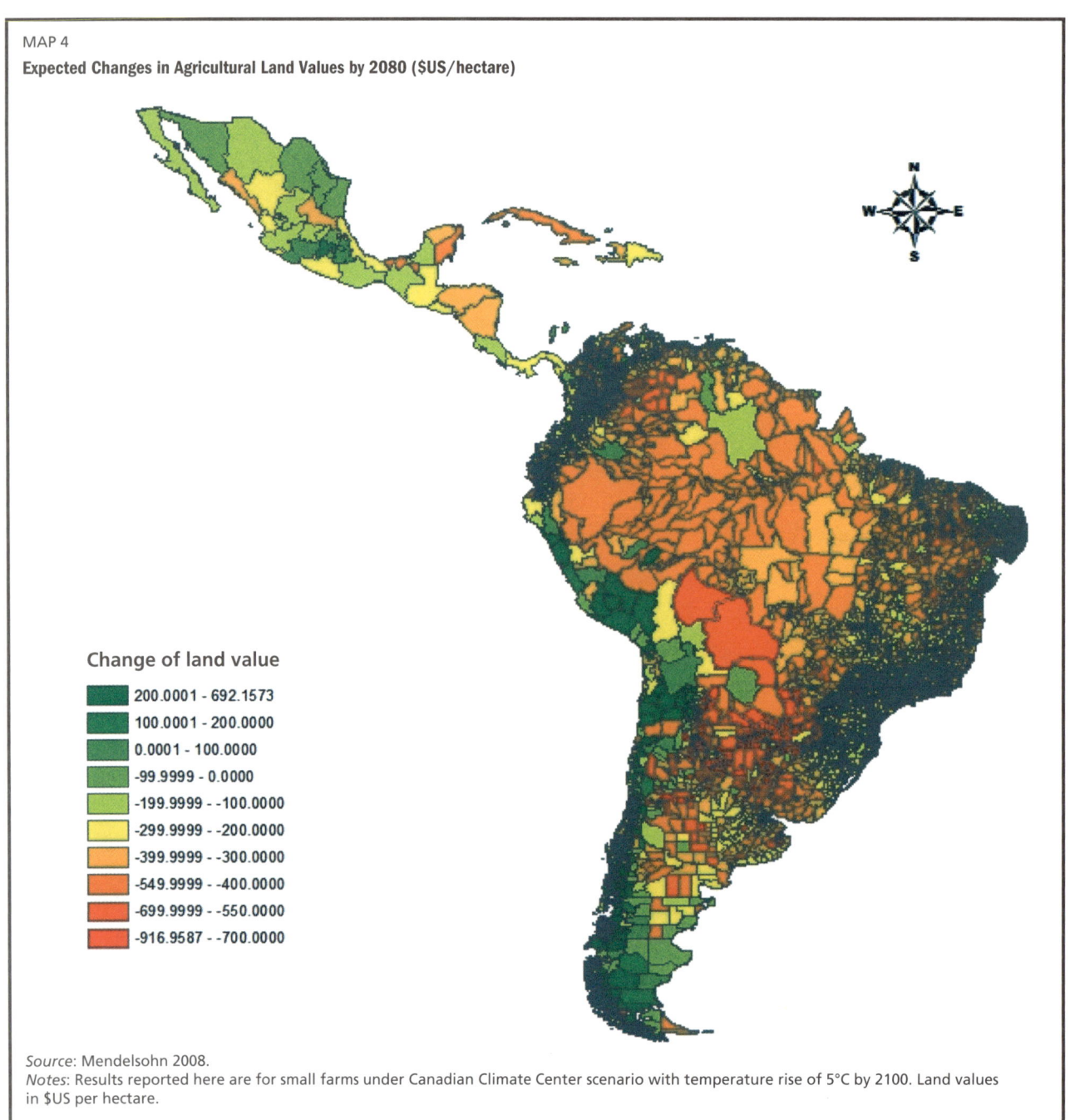

MAP 4

Expected Changes in Agricultural Land Values by 2080 ($US/hectare)

Change of land value

- 200.0001 - 692.1573
- 100.0001 - 200.0000
- 0.0001 - 100.0000
- -99.9999 - 0.0000
- -199.9999 - -100.0000
- -299.9999 - -200.0000
- -399.9999 - -300.0000
- -549.9999 - -400.0000
- -699.9999 - -550.0000
- -916.9587 - -700.0000

Source: Mendelsohn 2008.
Notes: Results reported here are for small farms under Canadian Climate Center scenario with temperature rise of 5°C by 2100. Land values in $US per hectare.

some areas on the Pacific and in the south of the continent showing possible gains.

What does this mean in terms of aggregate impact on GDP? For LAC as a whole, the agricultural sector is a small part of the economy, and following the pattern of almost all countries' historical experience, its share is expected to shrink further as the economies develop. The large impacts on agriculture translate into losses that are not very large relative to the econ-

omy as a whole. Past modeling efforts for Latin America have estimated agricultural losses to range from US$35.1 billion per year (out of US$49.0 billion total losses for all sectors, representing 0.23 percent of GDP),[25] to US$120 billion per year (out of US$122 billion total losses, 0.56 percent of GDP)[26] by 2100. A very recent study, based on a global general equilibrium model with endogenously determined emissions levels, projects total losses in LAC of around US$91 billion (about 1 percent of GDP) by 2050 if warming reaches about 1.79°C relative to 1900.[27] Since this is a permanent reduction in level of income, it would be equivalent in present value terms to a one-time shock of around 18.2 percent of GDP, using a discount rate of 5.5 percent.[28] None of these estimates include damage to noneconomic sectors, for example to ecosystems. Also, they do not take into account the possibility of increased frequency or potency of natural disasters, nor do they account for the possibility of catastrophic climate change from events such as the collapse of major ice sheets or melting permafrost.

What would be the impact of the expected changes in agricultural productivity on rural poverty? Answering this question requires modeling the way in which households would respond. In particular, the evidence suggests that there would be big differences in impact, depending on the degree of households' economic mobility. In the case of Brazil, for example, simulations based on municipal data suggest an average reduction of 18 percent in agricultural productivity by the middle of the century, which in turn could increase rural poverty by between 2 and 3.2 percentage points, depending on whether households are able or not to migrate in response to climate impacts. In either case, the effect of climate change is highly region-specific, depending on the regional changes in the climate per se, as well as the variation in productivity responses—which vary from increases of 15 percent to reductions of 40 percent in different parts of Brazil—and off-farm economic opportunities (map 5).

Economic damage from hurricanes and tropical storms is also likely to increase. Although there is no scientific consensus that hurricanes will become more common in the future, there is greater consensus that global warming is likely to cause their intensification. Indeed, global tropical storm intensity data since 1970 indicate an average increase in intensity of 6 percent for each increase of 1°F in sea surface temperature (Curry et al. 2008). Based on this kind of data, storm activity can be forecast using projections of the warming likely in the future. Such forecasts can take into account the influence of both natural variability and cycles as well as global warming on tropical storm frequency, intensity, and tracks.

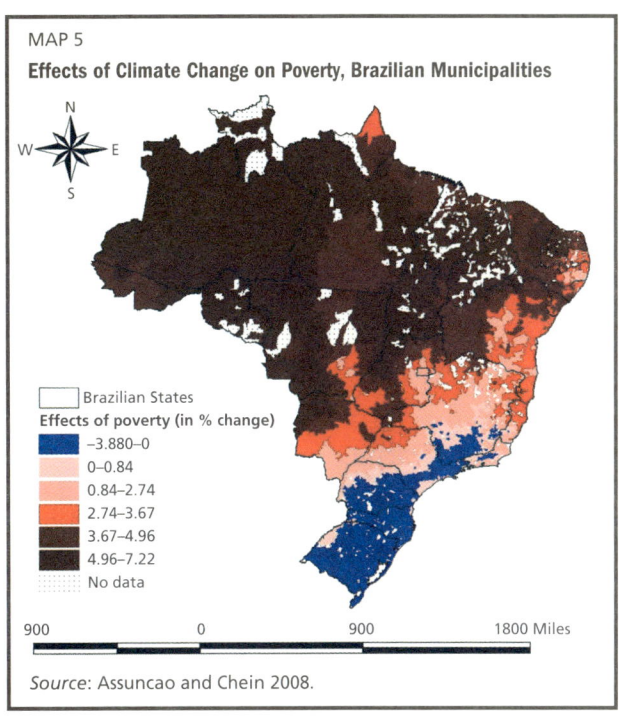

MAP 5

Effects of Climate Change on Poverty, Brazilian Municipalities

Brazilian States
Effects of poverty (in % change)
- −3.880–0
- 0–0.84
- 0.84–2.74
- 2.74–3.67
- 3.67–4.96
- 4.96–7.22
- No data

900   0   900   1800 Miles

Source: Assuncao and Chein 2008.

When this approach is used to model likely landfalls of tropical storms for Mexico's Gulf Coast, Central America, and the Caribbean region,[29] the projections indicate on average a very large increase in damage during the next 20 years, driven not only by greater storm intensity and, to a lesser extent, frequency (under two of the four scenarios modeled), but also by the increasing value of assets at risk resulting from economic development. In particular, estimates suggest a 10-fold increase in losses from hurricanes in Mexico's Gulf Coast during 2020–25, compared to the average five-year period during 1979–2006 (table 1).

Central America and the Caribbean would experience respectively threefold and fourfold increases over the same periods. In relative terms, Caribbean countries would still be the most affected, with cumulative losses of more than 50 percent of annual GDP by 2020–25, compared to about 10 percent of GDP for Mexico and 6 percent for Central America. Another recent study of the annual economic damages to 20 CARICOM countries circa 2080 from hurricanes and other natural disasters estimates these losses at US$4.9 billion in 2007 dollars, or about 5 percent of GDP per year (Toba 2008a; complete table of damages from all sources in annex 2 to this document).

11

TABLE 1

Cumulative Losses from Tropical Cyclones, Historic and Projected (millions of 2007 US$)

Country/region	Historic loss per 5 years (1979–2006)	Average losses (across 4 scenarios) per 5 years (2020–25)
Mexico	8,762	91,298
Central America	2,321	6,303
Greater Antilles	6,670	28,037
Lesser Antilles	925	2,223
Total	18,678	127,861

Source: Authors' calculations from Curry et al. 2008. Numbers reported are averages of the four scenarios considered.

The expected *disappearance of tropical glaciers in the Andes will have economic consequences on water and hydropower availability*. Modeling work and projections indicate that many of the lower-altitude glaciers in the cordillera could completely disappear during the next 10-20 years (Bradley et al. 2006; Ramírez et al. 2001). The Chacaltaya Glacier (see fig. 2), for example, may completely melt by 2013 (Francou et al. 2003).

Andean countries are highly dependent on hydropower (more than 50 percent of electricity supply in Ecuador, 70 percent in Bolivia, and 68 percent in Peru). Some of the hydropower plants depend in part on water from glacial runoff, particularly during the dry season. While the glaciers are melting, flows are high, increasing the threat of flooding. But this is a temporary phenomenon. Although it will continue for decades, eventually the volume of melt water will decline. This will create adjustment problems, as populations may have become dependent on the temporarily higher flows. In the longer term, while the disappearance of the glaciers might not affect total water supply (compared to the situation before glaciers began to melt), seasonal flow patterns are likely to change. Any reduction in the regulation of water flows in the dry season, caused by either increases in the variability of precipitation or reductions of natural water storage (glaciers, paramos, mountain lakes) would require new investments in reservoirs to maintain generation capacity. The phenomenon of glacier melt will also have serious consequences for water supply to the Andean cities.

Rising sea levels will economically damage coastal areas in numerous ways. With rising sea level, livelihoods, socioeconomic infrastructure, and biodiversity in low-lying areas of Mexico, Central America, and the Caribbean will be affected by increased salinity in coastal lagoons, such as Mexico's Laguna Madre. Saline intrusion from sea-level rise, combined with the above-noted reduced precipitation in the Gulf Coast region of Mexico, will cause increasing damage to wetlands there, reducing the many environmental services they provide. Agriculture could also be impacted by sea-level rise, particularly through loss of perennial crops, such as forests and banana trees, caused by the washing out of arable land and increased soil salinity (UNFCCC 2006).

It is very hard to value ecosystem services, and existing studies of the damage from sea-level rise have focused on more direct effects on economic activities, finding that these costs would be significant in vulnerable areas. Annual economic damage from climate change in CARICOM countries has been estimated at around US$11 billion by 2080, or 11 percent of GDP, with about 17 percent of the losses (around 1.9 percent of GDP per year) due to the specific effects of sea-level rise—loss of land, tourism infrastructure, housing, buildings, and other infrastructure.[30] In the LAC Region as a whole, estimates of total economic damages from sea-level rise range from 0.54 percent of GDP for a 1 meter rise to 2.38 percent for a 5-meter rise (Dasgupta et al. 2007), with the magnitude of losses differing greatly among the Region's countries (figure 4). These estimates are considered conservative, since they include only inundation zones, do not include damage from storm surges, and use existing patterns of development and land use.

Continued warming of sea-surface temperatures will cause more frequent bleaching and eventual dieback of the coral reefs, with high economic costs to the Caribbean. Future impacts of warming on the Caribbean reefs have recently been modeled, and the prospects are poor. With the IPCC's business-as-usual scenario (and a low temperature sensitivity scenario), the model predicts the mortality of all corals in the area between

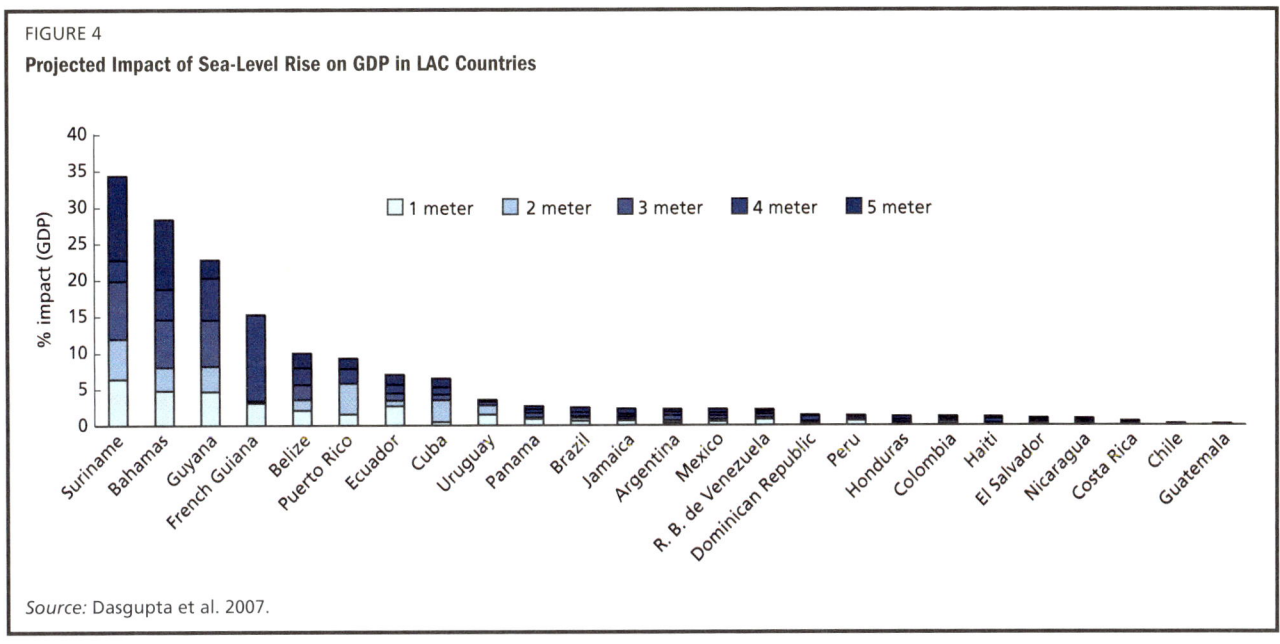

FIGURE 4

Projected Impact of Sea-Level Rise on GDP in LAC Countries

Source: Dasgupta et al. 2007.

2060 and 2070. Other scenarios assuming higher warming suggest that complete mortality could happen as soon as 2050. The model predicts that corals in the northern Caribbean are likely to suffer the impacts sooner than in more southern areas.

In addition to loss of biodiversity, this would have large direct socioeconomic impacts. Corals provide a natural protection against storm surges; as they bleach, the reefs disintegrate and thus eliminate this protection. As mentioned, around 65 percent of all species in the Caribbean depend to some extent on coral reefs, so the collapse of these reefs may have widespread impacts on fisheries as well as the ecologies of the area. Reefs are also a tourism attraction and as these bleach and disintegrate, they lose any esthetic value. These economic losses are inherently difficult to monetize, but table 2 presents estimates of their value in the event that 50 percent of coral reefs are lost. They suggest that total losses could range from 6 to 8 percent of the GDP of the smaller affected countries—including Belize, Honduras, and the West Indies.[31]

While forecasts of changes in local patterns of rainfall from global climate models are not as consistent as those of changes in temperatures, forecasts of major changes in some areas are fairly consistent. In arid and semi-arid regions of Argentina, northeast Brazil, northern Mexico, and Chile, *further reductions in rainfall could create severe water shortages.* The number of persons in Latin America living in water-stressed watersheds in 1995 was estimated at around 22 million. Modeling the effects of climate change, under the scenarios considered by the IPCC (Special Report on Emission Scenarios, 2001), by 2055, the number living in water-stressed areas in LAC would increase under three of the four scenarios by between 6 and 20 million persons (Arnell 2004). The economic consequences of such severe water shortages in the region

TABLE 2

Potential Value of Lost Economic Services of Coral Reefs, circa 2040–60 in 2008 US$ million (assuming 50% of corals in the Caribbean are lost)

	Low estimates	High estimates
Coastal protection	438	1,376
Tourism	541	1,313
Fisheries	195	319
Biodiversity	14	19
Pharmaceutical uses	3,651	3,651
Total	4,838	6,678

Source: Vergara, Toba, et. al. 2008b.

have not yet been analyzed, but could be large, particularly as they may lead to significant changes in the hydroelectric generation potential of the region, either in overall capacity or in its location.

Climate change is also likely to have multiple *impacts on health*, but the relationship is complex. Worldwide, the single most significant impact identified by the IPCC is an increase in malnutrition, particularly in low-income countries (Confalonieri et al. 2007), with mortality and morbidity from extreme events in second place. Other impacts identified include increases in cardiorespiratory diseases from reduction in air quality (due, for example, to more forest fires), changes in temperature-related health impacts (increasing heat stress, but reduction in cold-related illness, depending on the region), and an increase in water-borne disease if sewage systems become overloaded from heavy rainfall and dump raw sewage into sources of drinking water.

Of special concern in LAC will be the effects on malaria—mainly in rural areas—and dengue in urban areas. Vectors and parasites have optimal temperature ranges, and because mosquitoes require standing water to breed, changes in precipitation are also expected to have an effect on the prevalence of these diseases. In areas that are now too cool for such vectors to survive, higher temperatures could allow expansion both of the range and of the seasonal window of transmission. In areas where temperatures are now close to the upper threshold of tolerance, the range could contract. Areas with higher precipitation will have an increased risk. In Colombia, there is evidence that temperature is important for dengue transmission,

while increased precipitation is a significant variable contributing to malaria transmission. An increase in the number of cases of malaria in Colombia has already been observed, from about 400 per 100,000 in the 1970s to about 800 per 100,000 in the 1990s. Based on statistical models of the incidence of both malaria and dengue, and forecasts of change in precipitation and temperatures (derived from eight global circulation models used in the fourth assessment of the IPCC), the total number of dengue victims is forecast to increase by around 21 percent by 2050 and by 64 percent by 2100. Similarly, the incidence of malaria is expected to increase by 8 percent by 2050, and by 23 percent in 2100 (table 3).

It is worth noting that the corresponding economic costs, in terms of lost productivity and the cost of treating the additional victims, would be relatively small: US$2.5 million for the five-year period 2055–60, and US$7.5 million for the period 2105–10.[32] However, an important caveat in interpreting these results is that the additional cases were calculated only in the municipalities in which the corresponding disease was present in the 2000–05 period; the cost estimates above do not consider the potential spread to new municipalities.

On the other hand, areas receiving less rain may experience a reduction in malaria risk, as forecast for Central America and the Amazon.[33] But—underscoring the complexities in forecasting the net health impact of drier weather—the seasonal pattern of cholera outbreaks in the Amazon basin has been associated with lower river flow in the drier season.[34] No overall assessment has been carried out of the net

TABLE 3

Additional Numbers of Cases of Malaria and Dengue for 50- and 100-Year Future Scenarios

Vector-borne disease	Historic total number during the 2000–05 period	Additional number of cases for a 6-year period. 50-year scenario	Additional number of cases for a 6-year period. 100-year scenario
p. falciparum malaria	184,350	19,098	56,901
p. vivax malaria	274,513	16,247	48,207
Dengue	194,330	41,296	123,445
Total	653,193	76,641	228,553

Source: Blanco and Hernandez 2008.

health effects for the LAC region as a whole, but recent national health impact assessments in both Bolivia and Panama, for example, have concluded that on balance there is likely to be an increased risk of infectious disease in those countries.

3. The Need for a Coordinated, Effective, Efficient, and Equitable Global Response

The evidence presented so far indicates that climate change will impose significant costs on mankind and ecosystems. Attempts to minimize these damages can be broadly grouped into two classes. The first comprises efforts to *mitigate* climate change, which in the jargon of the climate literature means reducing GHG emissions so as to slow down global warming and other climate trends.[35] The second group of possible responses comprises so-called *adaptation* actions, aimed at adjusting natural or human systems in order to moderate harm or exploit beneficial opportunities associated with climatic stimuli or their effects. While there are many kinds of actions that provide significant cobenefits while helping to mitigate or to adapt to climate change, in general, investments in mitigation and adaptation have some costs. These costs may be incurred in the form of financial costs (for example, the additional cost of using wind power instead of coal to generate electricity), or as opportunity costs (for example, the income-generating opportunities forgone by preserving a forest). In order to determine what is the optimal global response to the climate change challenge, these costs must be weighed against the benefits of avoiding future damages.

The tradeoffs and synergies between mitigation and adaptation measures in principle call for an integrated approach to making simultaneous decisions on optimal levels of effort on both fronts.[36] But in a simplified framework, one can focus on the optimal level of mitigation efforts and assume that, given the resulting expected climate change impacts, adaptation expenditures will be decided optimally, by taking into account the corresponding costs and benefits of such actions.[37] Both the marginal costs and the marginal benefits of mitigating climate change depend on the scale of the emission reductions to be undertaken. On one hand, the costs of additional mitigation efforts tend to increase with the level of emission reductions. Low levels of emission reductions can be attained at relatively low costs; as reduction targets become more ambitious, these cheap solutions are exhausted and more expensive investments are required. The marginal benefits of mitigating climate change (the additional adaptation expenditures and residual damages avoided), on the other hand, tend to fall with the scale of emission reduction efforts.[38] The optimal degree of effort to mitigate the consequences of climate change would be the point at which the marginal cost of reducing emissions by one more ton just balances the damages avoided by doing so: Q* in figure 5, with a socially efficient price of carbon of P*. In a world in which all costs and benefits were taken into account by the same decision makers with perfect information, this optimal solution might be reached.

In practice, however, this outcome is unlikely for two reasons. First, emitters only absorb a very small fraction of the associated social costs, which are largely paid by others, most of whom belong to future generations. So individual agents—and countries—have an incentive to "free-ride" on the mitigation efforts of others. Moreover, even if some countries with large expected damages may decide to take mitigation actions unilaterally, the opportunities in these countries are not likely to be as cost-efficient as those in other countries.

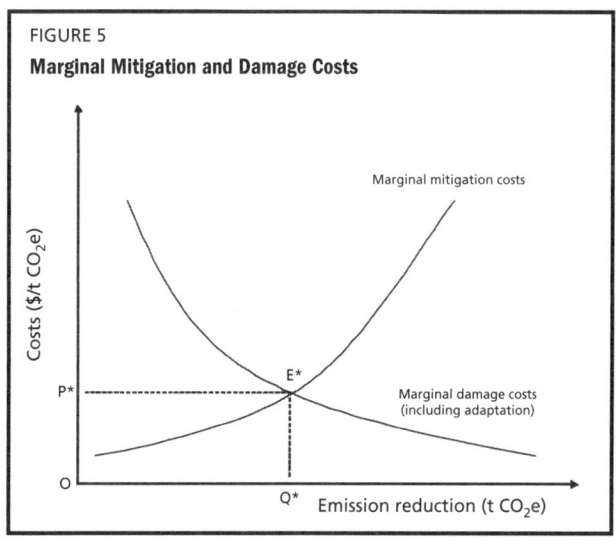

FIGURE 5
Marginal Mitigation and Damage Costs

Indeed, there is no reason to expect that countries with the highest risk exposure would also happen to have the lowest mitigation costs. In summary, global mitigation through uncoordinated individual efforts is likely to be (a) too small, (b) implemented too late, and (c) undertaken by the wrong countries.[39] In order to have any chance of reaching a level of mitigation and adaptation efforts close to that which would prevail in the absence of "free-riding," the world as a whole needs to come to a joint agreement.

But second, even with collective action, determining the optimal level of mitigation effort would be difficult because information required to estimate both the costs and the benefits is very imperfect. In particular, it is very hard to quantify the probabilities associated with specific climate impacts. In this regard, when dealing with climate change, policy makers are confronted not only with *risk*—randomness with known probabilities—but also with *uncertainty*.[40] The chain of causality between emissions today and the future impacts of climate change has many links, and there is a great deal of scientific uncertainty involved in moving from each one to the next.[41] This greatly complicates expected cost-benefit analyses. Moreover, there are potentially catastrophic climate impacts, the probability of which is thought to be low but is not well known. And the global climate system has a lot of inertia, creating long lags between changes in emissions and the impacts on natural systems, meaning that by the time it is discovered that a catastrophe is coming, it may be too late to avoid it. These considerations may make it prudent for policy makers to adopt an approach based on precaution, in which a large weight is assigned to the objective of avoiding such events.

In practice, this leads to a focus on establishing targets for GHG stocks, for which the probabilities of high levels of global warming with catastrophic consequences are estimated to be relatively small. This implicitly amounts to a willingness to pay an "uncertainty premium" so as to preempt those events. The definition of the specific targets that would shape public policies is akin to an iterative process of risk management, informed by the evolving scientific evidence on the sensitivity of climate to GHG concentrations, the damage costs from climate change, and the technological options for mitigation.

In fact, the 1992 agreement on the United Nations Framework Convention on Climate Change (UNFCCC), which has been ratified by 189 countries, explicitly recognizes as its overarching objective the stabilization of GHG concentrations at a level that avoids "dangerous" anthropogenic climate change. While there is as yet no universally accepted definition of such "dangerous climate change," one approach is to focus on reducing the prospect of encountering biological and geological "tipping points,"[42] when a system goes abruptly and irreversibly from one state to another, with wide systemic consequences, either for the world as a whole or for some regions. Examples would include the permanent loss of valuable ecosystems and/or species, and the possible disruption of key intrinsic processes of the climate system itself—for example, loss of the Amazon, the disintegration of the West Antarctic or the Greenland ice sheets. Some socioeconomic impacts could also be considered "dangerous" in the sense that if certain critical levels—for example, large cumulative socioeconomic impacts or serious disruptions of current practices—are reached, there could be consequences for human well-being that could be considered ethically or politically unacceptable (at least from a local perspective), or even produce large-scale social disorder. Examples could include levels of climate change that would trigger catastrophic food or water shortages, extensive coastal flooding, or the widespread dissemination of malaria or other tropical diseases

Avoiding "dangerous" impacts

As per the evidence presented above, the actions taken so far under the UNFCCC framework have not been bold enough to move the world away from potentially "dangerous" climate change trajectories.[43] What would it take, in terms of emission reductions, to avoid such paths? There is no single answer, but the more stringent the reductions, the lower are both the likelihood of catastrophic events and that of reaching "dangerous" levels of cumulative negative socioeconomic impacts. The most stringent potential targets considered by the IPCC call for stabilization of GHG concentrations within a range of 445 to 535 ppm CO_2e. The likely temperature increases associated with these targets are between 2°C and 2.8°C with respect to preindustrial levels. To achieve these targets global emissions would have to peak by 2020 at the latest. By 2050 they would have to drop to between 30 and 85 percent of the 2000 level. The costs of achieving these goals, based on 15 climate models considered by IPCC, is estimated to be a reduction of up to 3 percent of global GDP in 2030 and up to 5.5 percent by 2050.

An alternative set of targets considered by the IPCC would imply stabilizing GHG concentrations at levels between 535 and 590 ppm CO_2e. The cost of achieving these targets would be lower than for the more stringent targets mentioned above—up to 2.5 percent of global GDP in 2030 and 4 percent in 2050—but expected temperature increases would be slightly higher—between 2.8°C and 3.2°C.

Note, however, that given the large uncertainties involved, much higher rates of warming would still be possible (albeit improbable), even if the above targets were met. The expected level of global warming for the second group of targets, for example, could increase to almost 5°C if one were to use the more pessimistic available estimates (instead of the mode) for the so-called climate sensitivity parameter.[44] Similarly, Stern (2008) estimates that for a stabilization target of 550 CO_2e ppm there would be a 7 percent probability of temperature increases above 5°C, which could potentially lead to the melting of most of the world's ice and snow, as well as to sea-level rises of 10 meters or more, and losses of more than 50 percent of current species.

Effectiveness and efficiency call for developing country participation

Because of the scale of the emission reductions that are required, an effective global agreement to mitigate climate change will necessarily have to involve both industrialized and developing countries. This is the result of the simple arithmetic of the situation. Assume, for example, that stabilization targets of 535 to 590 CO_2e ppm—one scenario considered by the IPCC—were to be adopted. On a per capita basis, and

for the world as a whole, emissions would have to be reduced from about 6.9 tCO_2e in 2000 to between 3.2 and 4.8 tCO_2e in 2050. Even if rich countries would agree to reduce their emissions by 100 percent (thus becoming "carbon-neutral"), these targets would be met only if developing countries were to reduce their per capita emissions by as much as 28 percent by 2050.[45]

Developing countries' participation, however, would be needed not only to guarantee effectiveness but also to ensure that stabilization targets are reached efficiently, that is, at the least possible global cost. Assume, for example, that by 2030 a global uniform price of carbon of US$100 per ton of CO_2e was the outcome of a global "carbon tax" or a "cap-and-trade" scheme. As shown by the IPCC, this would lead to sufficient emission reductions to stabilize GHG concentrations in the range of 445 to 535 ppm CO_2e.[46] While these mitigation investments would be spread across many sectors, in most of them (the only exception being transport) more than 50 percent of the global mitigation potential would be located in developing countries. In fact, in the cases of industry, agriculture, and forestry, almost 70 percent of the global potential for reducing emissions comprises opportunities in developing countries.[47]

Clearly, developing countries' engagement is indispensable if those targets are to be met, so strong incentives to become part of the solution are in everyone's best interest. This approach would ensure that the world takes advantage first of those mitigation opportunities that offer the largest "bang for the buck." In other words, a globally efficient solution is only possible if reductions take place in countries that have the greatest potential for low-cost reductions, not necessarily where emissions are the highest. The global savings from such an efficient solution would be large. A recent study, for example, finds that reducing global emissions by 55 percent in 2050 globally—relative to a baseline business-as-usual path—would cost 1.5 percent of global GDP using a uniform carbon tax. The same global emission reduction—implemented in such a way that each country cuts its own emissions by 55 percent—would cost 2.6 percent of global GDP, or about 73 percent more than when using the more efficient approach.[48]

The need for the global response to be equitable

Would a rapid and substantial contribution of developing countries to the funding of global efforts to mitigate climate change be compatible with equity considerations? Clearly not, for two reasons, which together are at the heart of the principle of common but differentiated responsibility established by the UNFCCC. First, developing countries already face the challenge of poverty reduction and are the most vulnerable and the least able to adapt to the adverse effects of climate change. They can hardly be expected to shoulder the additional burden of reducing their GHG emissions. An equitable solution would allow developing countries to attain the quality of life that has been achieved by the current developed nations over the last 100 years.

Second, industrialized countries carry a much larger historical responsibility for the existing GHG concentrations that are driving climate change. The lower level of responsibility of developing countries can be illustrated by the fact that the cumulative energy-related emissions of rich countries from 1850 to 2004 are, on a per capita basis, more than 12 times higher than those of developing countries—respectively 664 and 52 tCO_2 per capita.[49] Thus, even though their share of the world's population is only about 20 percent, industrialized countries are responsible for 75 percent of the world's cumulative energy-related CO_2 emissions since 1850. This leads many observers to conclude that rich countries should assume a much larger share of the cost that will be associated with reducing global GHG emissions.

The relatively small contribution to cumulative emissions of even some of the largest developing countries is illustrated in figure 6. It shows that emissions grew with income at much faster rates when today's rich countries were industrializing than has been observed in recent decades in China, India,

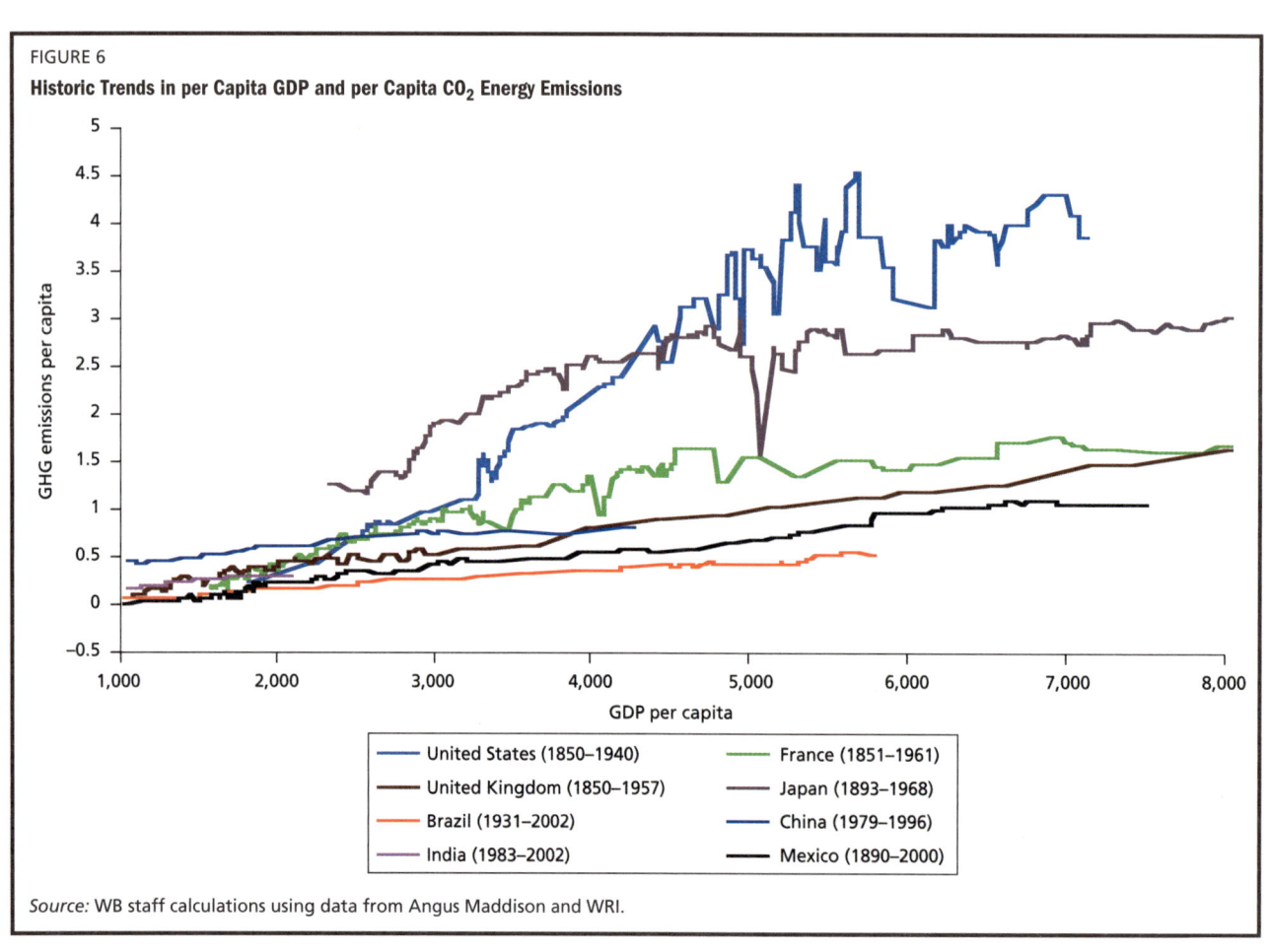

FIGURE 6

Historic Trends in per Capita GDP and per Capita CO_2 Energy Emissions

Legend:
- United States (1850–1940)
- United Kingdom (1850–1957)
- Brazil (1931–2002)
- India (1983–2002)
- France (1851–1961)
- Japan (1893–1968)
- China (1979–1996)
- Mexico (1890–2000)

Source: WB staff calculations using data from Angus Maddison and WRI.

Brazil, and Mexico. In other words, thanks to technological change, development has already become much less carbon-intensive than it was in the past.

Can an effective, efficient, and equitable global agreement be reached?

The discussion above implies three desirable characteristics for a coordinated response to the challenges of climate change. First, effectiveness in meeting stabilization targets that would likely serve to avoid "dangerous" impacts would require emission reductions to take place in both industrialized and developing countries.

Second, efficiency would require a mechanism to establish some kind of uniform price for carbon, so that the reductions would be carried out in the ways and places that it could be done most cheaply, and much of this will be in developing countries. Third, equity considerations would call for developed countries to carry a disproportionately larger share of the cost burden.

Is it possible to build a "global deal" that could satisfy both equity and efficiency considerations? The answer is in principle a clear yes, by decoupling the *cost* of mitigation from the *site* of mitigation (Spence et al. 2008), but the task will not be easy. The delinking could be achieved in several ways. One option is to adopt an international cap and trade scheme, through which a common price on carbon would emerge even if countries agreed on different levels of contributions to global efforts—that is, different caps on emissions. Resources would flow automatically to pay for emission reductions in countries that offer the lowest-cost mitigation opportunities, thus potentially funding an important level of mitigation efforts. A similar outcome could be achieved with a carbon tax mechanism—and some authors argue that such a mechanism might even be easier to negotiate and easier for developing countries to administer (Aldy et al. 2008). But with a carbon tax, equity would require a parallel agreement on a set of international resource transfers aimed at ensuring that the share of the global "bill" for climate change mitigation that is paid by each county is proportional to its responsibility for generating the problem and not necessarily to the country's actual contribution to its solution.

Considering the technical and political challenges associated with negotiating a global cap-and-trade scheme or a global carbon tax, however, it is worth considering other possible alternatives for decoupling the site of mitigation from its payment. While some of these alternatives may be more difficult to implement, some of them may constitute more acceptable outcomes from a political point of view. First, assuming that industrialized countries (including the United States) make deeper emission reduction commitments, expanded market-based instruments may play an important role. Second, complementary non-market financial instruments could help defray some of the costs of mitigation in developing countries, even if not serving to transfer emission rights to those who provide the funds. Finding the appropriate combination of these different types of instruments would be complex; it would have not only to adequately balance supply and demand within market mechanism(s), but also to balance, within the nonmarket mechanism(s), willingness to pay on the part of the industrialized countries and effectiveness to promote reductions in the south.

But if successfully negotiated, such a palette of climate finance instruments could bring all countries together into a common framework, and provide operational meaning to the phrase "common but differentiated responsibilities." In particular, a global agreement could confirm most (small) developing countries as continued hosts of scaled-up market-based mitigation efforts.

But it could at the same time provide the necessary incentives for the larger developing countries to gradually move toward adoption of their own climate mitigation commitments, which do not necessarily have to be Kyoto-type commitments. One example of how to alleviate the tradeoffs between economic development and climate change mitigation objectives would have some developing countries start with a focus on "climate-friendly" development policies, and transit over time, based on demonstrated capability (for example, as measured by per capita income) to commitments regarding the rates of growth of their emissions and, finally at some point in time, to some of them adopting emission reduction commitments (figure 7).

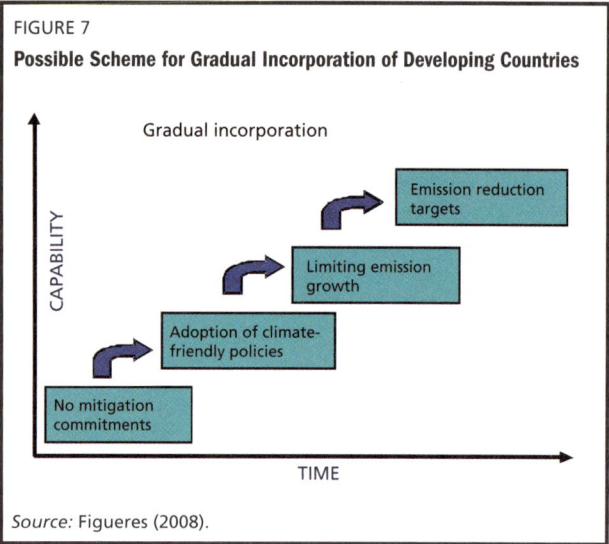

FIGURE 7

Possible Scheme for Gradual Incorporation of Developing Countries

Source: Figueres (2008).

In order to uphold the integrity of the system, all mitigation efforts, whether based on climate-friendly policies or eventually on targets, would have to be domestically measured and reported, and independently verified. In order to ensure fairness and equity, the gradual incorporation of developing countries could be linked to—that is, be conditional upon—industrialized countries' verified performance (in terms of both the provision of financing for developing countries mitigation efforts and emission reductions achieved at home).

Moreover, an agreement would have to be reached on possible objective criteria for defining the thresholds that would trigger an increasing degree of incorporation of developing countries. In this respect, it is important to recognize the wide variety of country circumstances that are found not only across rich and poor countries, but also within the group of developing countries. In this context, we now turn to an analysis of how the specificities of the Latin America and Caribbean Region may affect its participation in a global coordinated policy response to the climate change challenge.

4. LAC's Potential Contribution to Global Mitigation Efforts

There are many motivations for Latin American and Caribbean countries to participate actively in global efforts to mitigate climate change. However, one could divide those reasons in two groups. First, it is in the region's best interest to do so; thus, it *should* do it. Second, the region is well placed, in terms of its comparative advantages and potential to reduce GHG emissions, to make an important contribution to global efforts: therefore, one could argue that LAC *can* do it.

Why LAC should be "ahead of the pack"

As described above, LAC is already being hit by negative climate change impacts. If GHG emissions continue unabated, the Region is likely to suffer much more severe impacts in the future. As a result, LAC has a vested interest in the success of global mitigation efforts. While it is recognized that the challenge needs a global response, leadership on the part of LAC would have a clear positive effect. In addition, there are at least two types of instance in which undertaking its own climate mitigation efforts may involve benefits for the Region, even though it would contribute only modestly to avoiding future climate change damages, given the Region's relatively limited emissions.

First, in many cases emission reductions can be obtained while pursuing other economic development objectives. In these situations, which we will discuss in detail below, climate change mitigation would be a byproduct of actions that the region would be interested in pursuing anyway in order to promote sustainable growth and reduce poverty, regardless of climate change. Thus, one could argue that mitigation in these cases would involve "no regrets in the present." The main examples of such opportunities are related to investments aimed at increasing energy efficiency, reducing deforestation, improving public transportation, deploying renewable energy sources, developing low-cost and sustainable biofuels, increasing agricultural productivity, and improving waste management.

Second, climate mitigation may also involve "no regrets in the future" in a "carbon-constrained world," especially if the region takes a leadership position in the deployment of low-carbon technologies. In particular, given the growing scientific consensus regarding the real and present threats posed by climate change, developing as well as developed countries ultimately will have to take strong action to reduce GHG emissions. As a result, companies and countries will face an increasing pressure to internalize the social costs imposed by emissions.

Anticipating this shift has a number of advantages. Chief among them is the possibility of avoiding the "regrets" associated with the effect of future carbon taxes, emission caps, or other related regulations on the future profitability of current investments in "high-carbon" technologies, or the need to undertake large and rapid mitigation efforts later. These potential "regrets" could be minimized by taking into account early on, in the corresponding investment decisions, the prospective future emergence of carbon pricing. In other words, by incorporating expectations about the likelihood of future government policies and carbon market forces penalizing GHG emissions, companies and countries could improve the expected profitability of their investments, especially in "carbon-intensive" sectors.

Additional benefits of such an "early-mover" approach could be associated with the possibility of developing new comparative advantages in low-carbon technologies. This potential benefit would apply to companies and countries that make early investments in technologies for which market growth eventually accelerates as global mitigation efforts gain momentum. Finally, by moving "ahead of the pack," LAC countries that make early investments in low-carbon technologies are likely to benefit to a larger extent from international financing mechanisms. Indeed, the development and early deployment of low-carbon technologies is likely to benefit from some sort of subsidization, including through international financing mechanisms. By adopting an "early-mover" approach, LAC countries could thus be able to reduce the domestic costs of their investments in innovative low-carbon technologies.

It is worth noting, however, that there are also downside risks associated with being an early mover. First, the underlying assumption that the world will soon move to more aggressive limits on GHG emissions could be proven wrong. This could happen, for instance, if new scientific evidence appears that reduces the current sense of urgency with regard to climate change, or technological breakthroughs

reduce the need to abandon current production technologies. Second, it is possible that a global agreement with all the desirable characteristics discussed in the previous section will prove politically infeasible, at least in the short and medium terms, which would reduce the potential for international cost sharing of early actions. Third, the cost of low-carbon technologies will tend to fall over time, as a result of cumulative investments in research and development and dynamic economies of scale. Thus, there would be an advantage in waiting for adoption costs to fall, which would need to be weighed against the advantages of earlier action.

To deal with these risks, a prudent approach would involve focusing first on investments that involve clear "no regrets" in the present, and fewer technological uncertainties. The decision to move into riskier investments—with potential "no regrets" in the future—could then be conditional on the achievement of sufficient momentum in global mitigation efforts and/or to access to international cost-sharing mechanisms that would allow compensating for the risks described above. Besides minimizing the above-described downside risks associated with LAC being an "early mover," this approach would have the added advantage of helping create momentum toward a global agreement for addressing climate change challenges. Indeed, a strong show of leadership by medium-income countries such as those in LAC could help pave the road for increasing commitments among their high-income counterparts. In fact, this type of approach has already been adopted by a number of medium-income countries, both from LAC and other regions.[50]

LAC's potential for "no-regret" mitigation

As argued before, LAC has an interest to take the lead, among developing countries, in participating in international efforts to mitigate climate change. This section argues that the Region is also well placed to take such a leadership position. To that end, we first present some basic stylized facts on the levels and trends of LAC countries' GHG emissions and then proceed to documenting concrete "no-regrets" mitigation opportunities in various economic sectors.

LAC's GHG emissions: composition, levels, and trends

The first objective of this section is to identify the areas in which LAC's emissions are relatively low, thus suggesting that the Region has a comparative advantage for pursuing a low-carbon growth path. Second, we aim at characterizing those areas in which there appear to be opportunities for reducing the Region's emissions, as suggested either by large ratios of emissions to GDP or by high rates of emission growth. To achieve these goals we compare LAC's emission patterns with those of other regions of the world, and also explore the extent of heterogeneity existing across LAC countries.

The composition of LAC's GHG emissions

LAC has historically made a substantial contribution to keeping levels of atmospheric CO_2 low. First, LAC is host to about one-third of the world's forest bio-mass, and two-thirds of the biomass existing in tropical forests.[51] Were the large amounts of carbon stored in LAC's forests to be released to the atmosphere, current GHG concentrations would already be much higher. Second, LAC has enjoyed many decades of growth with very clean power. In particular, thanks to its low use of coal-fired plants and its large use of hydroelectricity, LAC's power sector generates 40 percent less CO_2 emissions per unit of energy than the world as a whole—74 percent less than China and India, and 50 percent less than the average for developing countries.[52]

Not surprisingly, the composition of LAC's flow of GHG is dominated by CO_2 emissions from land use change, which constitutes 46 percent of LAC's emissions, versus 17 percent for the world (figure 8). Put simply, because some other regions long ago cut down a large part of their forests, LAC has a large proportion of the trees that are still standing, and as a result it

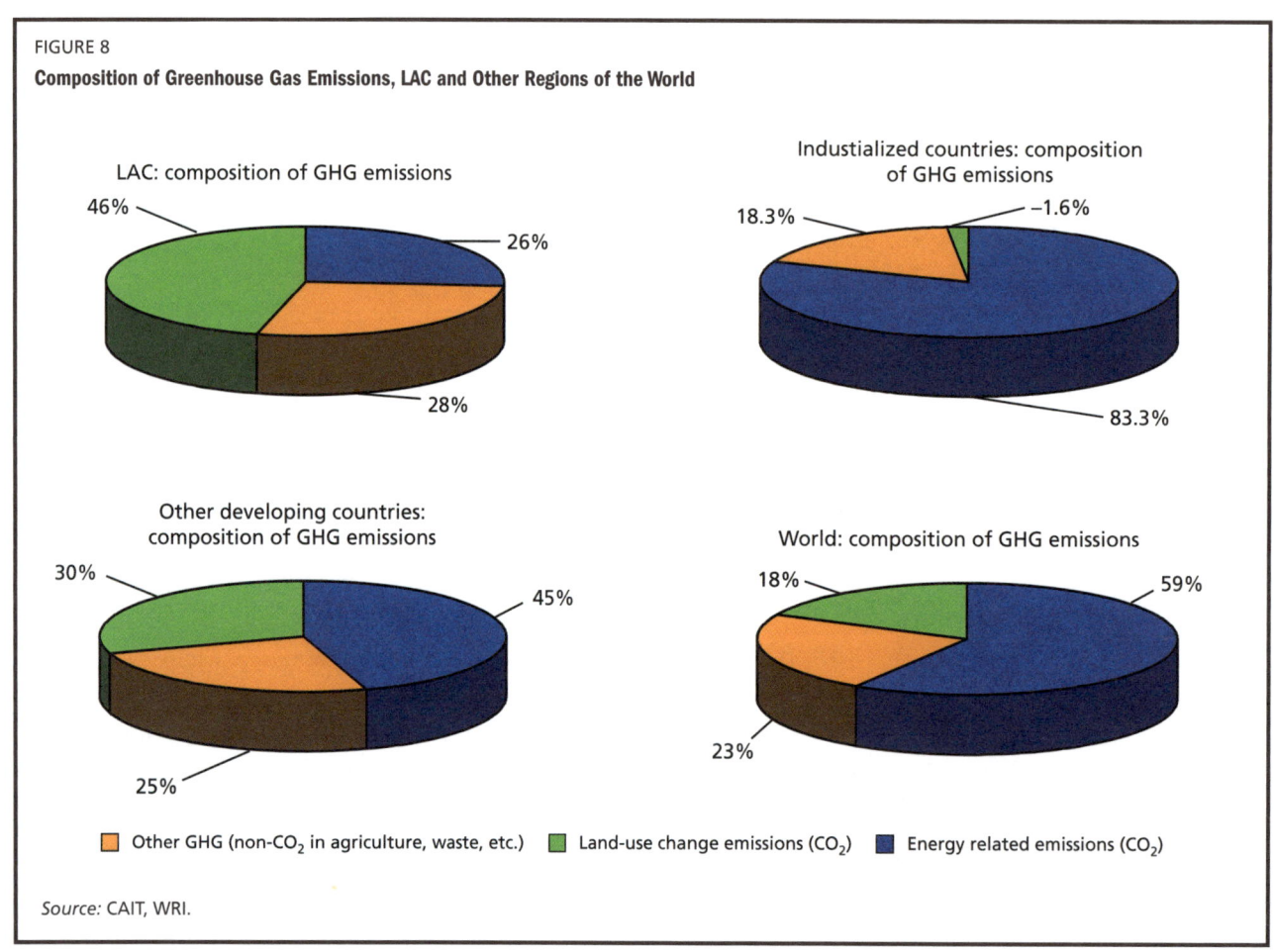

FIGURE 8

Composition of Greenhouse Gas Emissions, LAC and Other Regions of the World

LAC: composition of GHG emissions
46% · 26% · 28%

Industialized countries: composition of GHG emissions
18.3% · −1.6% · 83.3%

Other developing countries: composition of GHG emissions
30% · 45% · 25%

World: composition of GHG emissions
18% · 59% · 23%

■ Other GHG (non-CO_2 in agriculture, waste, etc.) ■ Land-use change emissions (CO_2) ■ Energy related emissions (CO_2)

Source: CAIT, WRI.

also has a large fraction of the emissions generated by cutting them. In contrast, the share of CO_2 energy emissions in LAC's total GHG emissions (26 percent) is much smaller than at the global level (59 percent). The remainder of LAC emissions (about 28 percent compared to 23 percent for the world as a whole) are other GHG generated mainly in the agricultural sector—70 percent in the case of LAC vs. 55 percent for the world—but also as a result of waste disposal as well as industrial and extractive activities.

These first basic traits of LAC emissions have a number of general implications in terms of identifying the main challenges, looking forward, for exploring the Region's mitigation potential. First, it is clear that LAC has an enormous mitigation potential associated with reducing land-use change emissions, which implies looking in detail at the potential for avoiding deforestation and implementing afforestation and reforestation projects. Second, it would be critical to maintain and further reduce LAC's relatively low ratio of emissions to energy, including emissions from power generation, transport, industrial activities, and commercial and residential buildings.

Of particular concern is the recent trend toward increasing the carbon intensity of power supply due to the shift away from hydroelectricity and toward natural gas and coal, a trend that is exacerbated in future projections of the sector. In order to at least maintain the past relatively low level of energy-related emissions, the Region would have to invest further in energy efficiency, renewables, and cleaner transport.

How large are the region's emissions?
LAC accounts for about 8.5 percent of the world's population and GDP, and for 12 percent of global emissions, considering all GHG. The Region's emissions are thus above the world average in terms of their ratio to both population and to GDP. While there is no agreement on how to measure responsibility and capability, those ratios could be used at least as indicative proxies for respectively the Region's *responsibility* and *potential* for reducing emissions.

On both counts, as shown in figure 9, LAC would be in an intermediate position, in between low- and high-income countries. Thus, LAC's per capita emis-

sions would be lower than those of industrialized countries, but higher than those of low-income. Figure 9 also shows that despite the large growth in GHG emissions observed in China and India during recent years, those countries still have much lower emissions per capita than LAC, and also a much lower ratio of emissions to GDP. Note, however, that if the focus is placed on energy emissions, LAC is among the regions of the world with lowest emissions per unit of GDP, and its emissions per capita are more than 30 percent below the world average

Is LAC moving in the wrong direction?
Over the past two and a half decades, energy emissions per capita have been relatively stable in LAC, while they have fallen in North America and Western Europe. A growth pattern similar to LAC's has been observed in Africa and Central and Eastern Europe. In contrast, the countries from Centrally Planned Asia (mainly China), the Far East (including India, South Korea, and Indonesia), and the Middle East have exhibited uninterrupted and explosive rates of growth in per capita emissions.

LAC's ratio of emissions to GDP has also remained relatively stable, experiencing only a 2 percent increase between 1980 and 2004. In contrast, there was a 28 percent decline in global emissions per unit of GDP during the same period, a 33 percent reduction in industrialized countries, and a 48 percent drop in the case of China and India. Other developing countries experienced relatively small declines: 9 percent in low-income countries and 4 percent in other middle-income countries (excluding LAC as well as China and India).

The fact that LAC's emissions per unit of output have remained relatively stable is to some extent surprising, given that the Region has achieved large reductions in the quantity of emissions per unit of energy consumed. In fact, this reduction in LAC's "carbon intensity of energy" has been almost totally compensated by a growing level of energy consumption per unit of GDP. As illustrated in figure 10, this is a trend that has only been observed in LAC and in low-income countries.[53] Indeed, during the same period, other middle-income countries (including

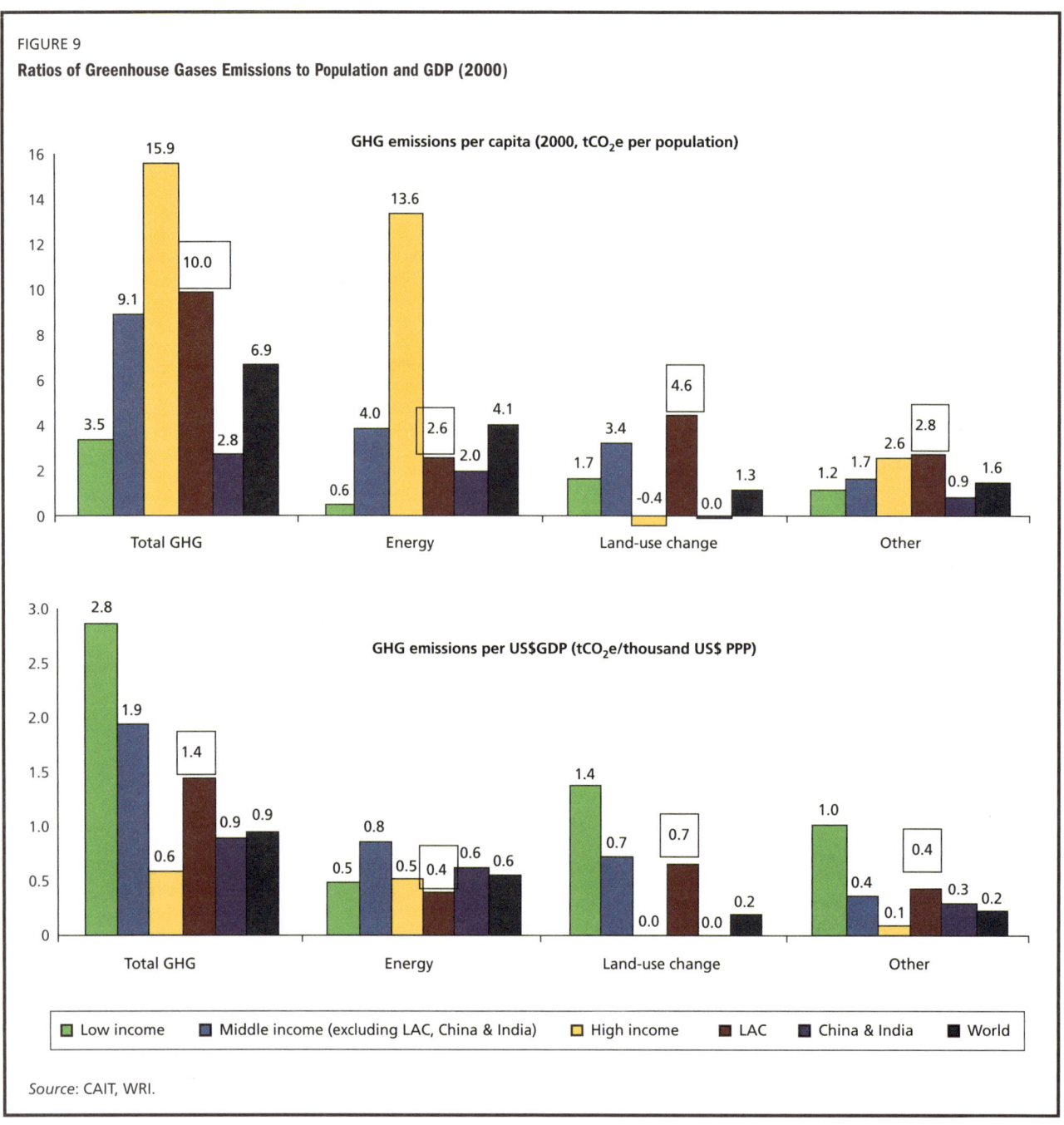

FIGURE 9

Ratios of Greenhouse Gases Emissions to Population and GDP (2000)

Source: CAIT, WRI.

China and India), as well as high-income countries, exhibited decreasing levels of energy intensity, especially in the years immediately following the oil shocks of the 1970s.

The good news is that most of the increase in LAC's energy intensity took place during the 1980s, and some significant reductions have already been observed since 2000. The bad news is that one of the main factors that is likely to have driven LAC's limited reaction to the increases in international oil prices of the 1970s remains largely unchanged.[54] Indeed, as explored in detail further below, energy prices in the Region continue to be heavily regulated in such a way that international price increases are only partially passed through to consumers and thus fail to provide the appropriate incentives to reduce consumption.

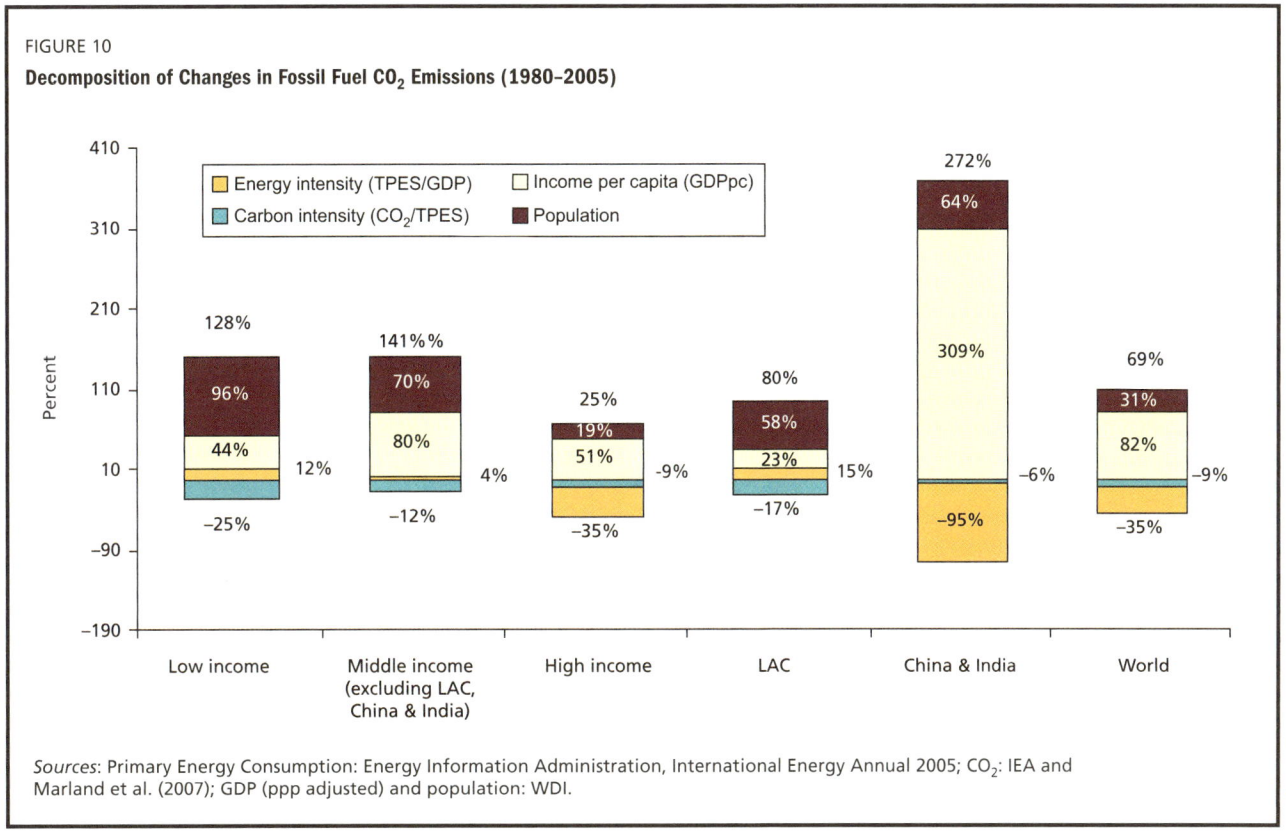

FIGURE 10

Decomposition of Changes in Fossil Fuel CO₂ Emissions (1980–2005)

Sources: Primary Energy Consumption: Energy Information Administration, International Energy Annual 2005; CO₂: IEA and Marland et al. (2007); GDP (ppp adjusted) and population: WDI.

Looking forward, the International Energy Agency (IEA) predicts that LAC's per capita energy-related emissions will grow by 10 percent between 2005 and 2015, and by 33 percent during 2005–30. These projections are much lower than those made for other developing countries—for example, energy emissions in China and India are expected to grow by more than 100 percent on a per capita basis between 2005 and 2030. However, LAC emissions are predicted to grow by more than the world average after 2015. While the IEA does expect significant reductions in LAC's energy intensity, it predicts no significant contributions to emission reductions in the Region to come from further declines in the carbon intensity of its energy. This is to some extent surprising, given that, as discussed below, LAC still has a very large potential for developing clean energy sources.

Cross-country differences in emissions patterns

About 85 percent of the Region's emissions are concentrated in six countries. Brazil and Mexico account for almost 60 percent of both the Region's total GHG

emissions and its GDP. Another 25 percent of LAC's emissions and GDP is accounted for by Argentina, Colombia, Peru, and República Bolivariana de Venezuela. A similar ranking emerges if one excludes emissions from land-use change, with the exception of Brazil and Mexico, for which the share of LAC total emissions respectively falls from 46 to 34 percent and increases from 13 to 21 percent.

While emissions from land-use change are responsible for almost half of LAC's total GHG emissions, their share varies widely across countries in the region. In five countries—Bolivia, Brazil, Ecuador, Guatemala, and Peru—land-use change accounts for at least about 60 percent of total GHG emissions. In contrast, in Mexico, Chile, and Argentina, the share of land-use change emissions is close to 15 percent. Brazil alone is responsible for 58 percent of LAC emissions from land-use change, followed by Peru with 8 percent, and by República Bolivariana de Venezuela and Colombia with about 5 percent each.

There is considerable heterogeneity across LAC countries in levels of GHG emissions, both in per

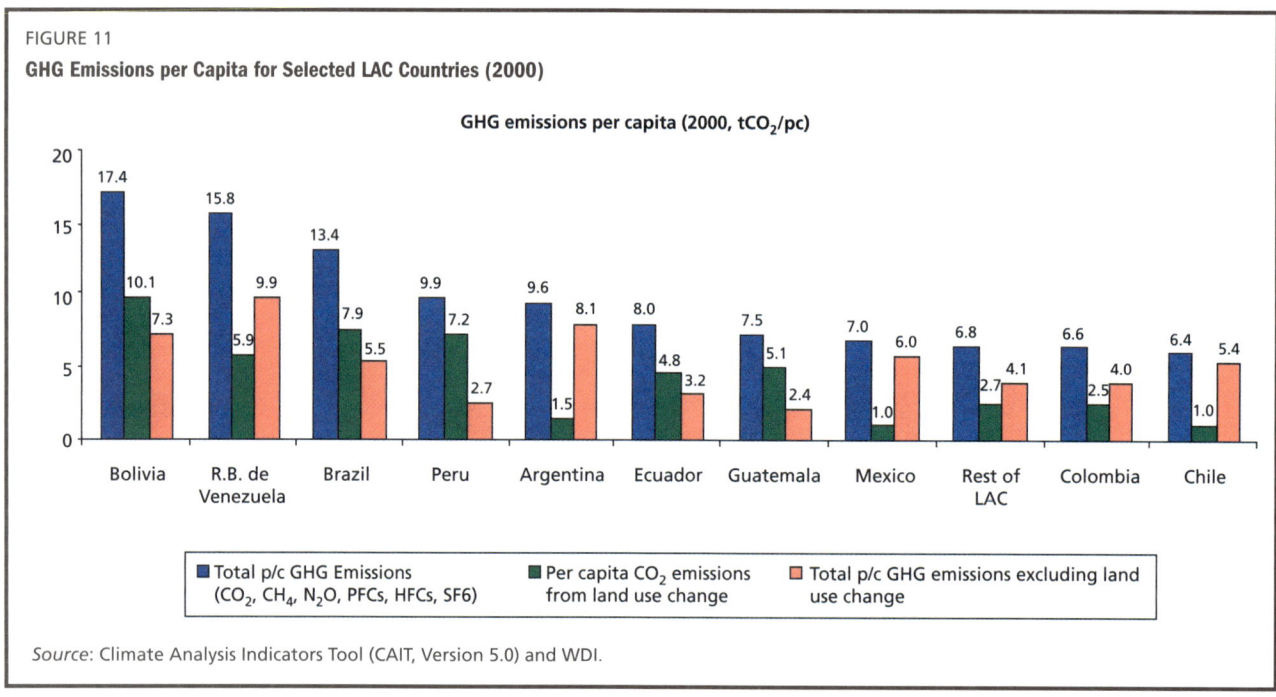

FIGURE 11

GHG Emissions per Capita for Selected LAC Countries (2000)

GHG emissions per capita (2000, tCO₂/pc)

Legend:
- ■ Total p/c GHG Emissions (CO₂, CH₄, N₂O, PFCs, HFCs, SF6)
- ■ Per capita CO₂ emissions from land use change
- ■ Total p/c GHG emissions excluding land use change

Source: Climate Analysis Indicators Tool (CAIT, Version 5.0) and WDI.

capita terms (figure 11) and as a ratio to GDP (figure 12). For instance, total GHG emissions per capita are between 13 and 17 tCO₂ per capita in Bolivia, República Bolivariana de Venezuela, and Brazil, and below 7 tCO₂ per capita in Chile, Colombia, and Mexico. The former three countries are also among the Region's top per capita emitters even if land-use change is excluded, although in this case their emissions per capita are much closer to those of Argentina, Chile, and Mexico.

The ratio of emissions to GDP and the rate of growth of emissions are possible measures of countries' mitigation potential. Indeed, where both of those variables are low, there is arguably little room for further emission reductions. Figure 12 exhibits the values of those two variables—the ratio to GDP in the horizontal axis and the emission growth rate in the vertical one—together with the absolute value of total emissions (size of the "bubble"). The top panel focuses on energy-related emissions and the bottom panel on land-use change (LUC) and non-CO₂ emissions (for example, from agriculture). In both cases, the point where the axes cross corresponds to the typical LAC country. Figure 12 suggests that some LAC countries have a relatively high mitigation potential in energy (for example, Argentina, Chile, Mexico, and

República Bolivariana de Venezuela), while for others the potential for reducing GHG emissions lies mainly in LUC or agriculture (for example, Brazil and Peru). A finer analysis of relative mitigation potentials for more disaggregated categories of emissions is reported in annex 1.[55]

How LAC can be part of the solution:
Specific "no-regrets" mitigation opportunities

As described above, LAC clearly has a comparative advantage in pursuing a low-carbon growth path, by means of implementing policies and programs to conserve its large forests and to maintain its relatively clean energy matrix. To realize this potential requires identifying concrete opportunities for reducing GHG emissions without compromising sustainable development objectives. As documented below, there are many ways in which the Region's emissions can be reduced at low cost, while at the same time reaping sizable development cobenefits. In some cases, these cobenefits have a value that would more than offset the costs of undertaking the measures; that is, there would be negative net costs. These could be called "no-regrets" options, in the sense that even if reducing emissions is not a consideration; a country should have "no regrets" in undertaking them, since they are good

FIGURE 12

GHG Emissions Growth and Ratio to GDP

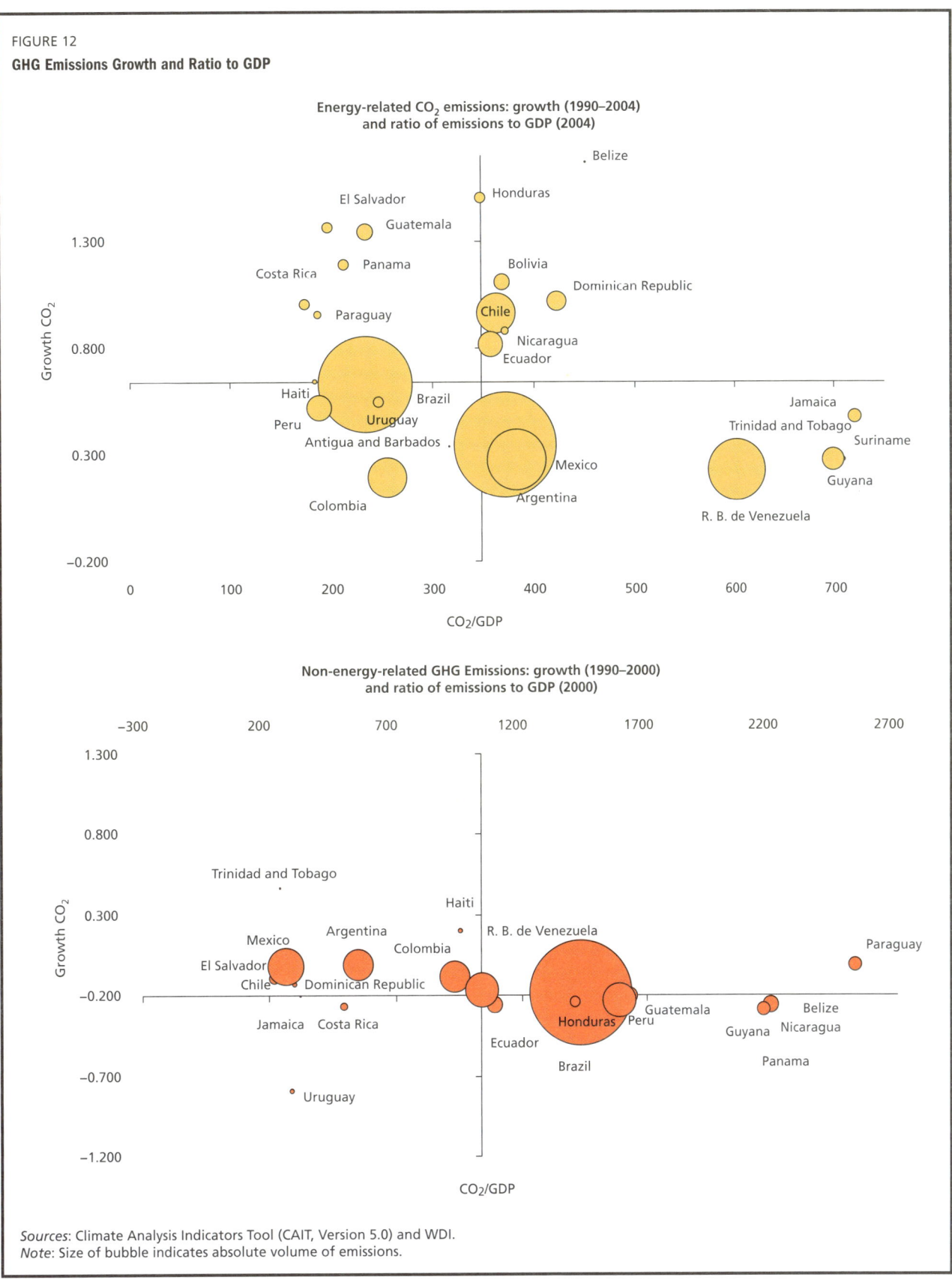

Sources: Climate Analysis Indicators Tool (CAIT, Version 5.0) and WDI.
Note: Size of bubble indicates absolute volume of emissions.

development policy. Where the cobenefits are financial, the negative net cost is reflected in pecuniary savings. Of course, the fact that these "low-hanging fruits" have not yet been harvested suggests that there are various obstacles—pecuniary or nonpecuniary. Concrete measures to address these barriers are discussed in section 5 of this paper.

Energy efficiency

Improving energy efficiency has important benefits beyond climate change mitigation. They include the ability to reduce energy demand in the short term, delay construction of new electric generating capacity, increase competitiveness by lowering production costs, and reduce fossil fuel consumption and the emission of local pollutants. Energy efficiency is particularly important for countries facing energy supply constraints as it can reduce the growth in demand in the near term, which avoids the administrative and legal processes and time needed for planning, licensing, and constructing new generating capacity.

By any measure, there is substantial untapped energy efficiency potential worldwide and in Latin America that could reduce GHG emissions at a relatively low or even negative cost. The IPCC calculates that about 25 percent of the global mitigation potential for carbon prices of up to US$100/tCO$_2$e could be achieved at negative social costs. About 80 percent of these no-regrets mitigation alternatives are associated with increases in energy efficiency in commercial and residential buildings. Similarly, the International Energy Agency estimates that energy efficiency accounts for more than half of the global energy-related emission abatement potential achievable within the next 20–40 years.[56]

In LAC, a recent analysis by the InterAmerican Development Bank estimates that energy consumption could be reduced by 10 percent over the next decade by investing in energy efficiency. The cost of such measures would be US$37 billion less than investing in new electricity generation capacity.[57] In the case of Mexico, ongoing studies sponsored by the World Bank suggest that between 2008 and 2030, GHG emissions could be reduced by about 15 million tons (Mt) of CO$_2$e through an increased use of cogen-

eration in the steel and cement industries and by means of efficiency improvements in residential and commercial lighting. In both cases the cost of achieving the corresponding emission reductions would be negative. The electricity savings from using more energy efficient lighting would amount to 6 percent of total generation in 2006, which would allow investments of about US$1.5 billion to be deferred, and saving US$1.7 billion in energy subsidies.

Additional opportunities for no-regrets investments have been identified in several recent studies. One study for Mexico found good opportunities for efficiency improvement in the residential, industrial, and public sectors.[58] Similar studies sponsored by the energy company Endesa in Argentina, Chile, Colombia, and Peru also suggest a large potential for emission reductions at negative costs in the area of energy efficiency.[59] In the case of Chile the largest potential is found in efficiency improvements in electricity generation, followed by improvements in the industrial and mining sectors. The studies for Argentina and Colombia find a sizable mitigation potential in the areas of residential and commercial lighting, while the Peru study found a large potential for energy efficiency improvements in the industry and agroindustry sectors.

Forestry

Efforts to harness the climate change mitigation potential of land-use change at the global level are focused on reducing emissions from deforestation and forest degradation (REDD) and, to a lesser extent, around afforestation and reforestation (A/R) activities. In addition to helping reduce net GHG emissions, forest conservation efforts also play important roles in supporting sustainable development in the corresponding areas, as well as in helping ecosystems and communities adapt to climate change.

In particular, forest conservation efforts can foster climate-resilient sustainable development by helping regulate hydrological flows, restore soil fertility, reduce erosion, protect biodiversity, and increase the supply of timber and nontimber forest products.[60] This is not to say that tradeoffs between mitigation and adaptation do not arise in A/R and REDD activities. There are, for example, documented cases of competition between tree plantation and agriculture in terms of the land and water that are needed, especially in arid and semi-arid regions.

Assessing the mitigation potential of A/R and REDD activities requires estimating land availability and the potential carbon sequestration or retention potential of the available land. The latter depends mostly on biophysical considerations (soil type, precipitation, altitude, and so forth) and the type of vegetation. Based on a literature review of regional bottom-up models, the IPCC estimates that the economically feasible potential of forestry activities in the LAC Region by 2040 ranges from 500 to 1,750 $MtCO_2$ per year, assuming a price of US$20/$tCO_2$. In particular, land available for A/R activities in LAC is estimated at 3.4 million square kilometers, most of it in Brazil. Other countries—especially Uruguay and some Caribbean countries—also offer a significant potential, at least in terms of the share of their corresponding territory.[61]

Empirical assessments of mitigation potential through REDD have focused on calculating the opportunity cost of avoided deforestation or, in other words, on the forgone income associated with conserving forests as opposed to implementing other economic activities in the corresponding land. To that end three different approaches have been used: local/regional empirical studies, global empirical studies (for example, those reported in the Stern Review), and global simulation models.[62] The results of a review of 23 different local models suggest a cost of avoided emissions from deforestation ranging from zero to US$14/$tCO_2$, with a mean value of US$2.51/$tCO_2$.

In comparison, the Stern Review estimated that deforestation could be reduced by 46 percent (in area terms) for a cost U$1.74–5.22 per tCO_2 with a midpoint that is 38 percent higher than the mean value of the estimates of local studies. Global models result in the highest cost per ton of avoided emissions, with values in a range of U$6–18/$tCO_2$ for reducing deforestation by 46 percent also. The large differences across models are driven by the selection of baselines (rate of deforestation based on past or expected deforestation rates), the assumptions about the carbon con-

tent of the forest, and the dynamics of the different variables and sectors considered (from static to global equilibrium models).[63]

Other relevant factors that will have an impact on the cost of REDD—beyond the opportunity costs discussed above—include costs related to the implementation of the corresponding government policies (for example, forest monitoring and regulation enforcement). Moreover, even when government policies focus on compensating stakeholders for conserving forest land, the costs of the corresponding programs may vary depending on whether the authorities price-discriminate between lands with different opportunity costs. Finally, one should also consider the fact that the activities forgone for the purpose of forest conservation may have not only private but also public benefits (taxes paid by logging companies to the government, loss of income as a result of unemployment, and so forth).

It is clear that further research is needed to improve our estimates both of the opportunity costs of avoiding deforestation and of the costs of implementing REDD policies. To assist countries in understanding how land-use change affects GHG emissions, and to tailor respective policy responses, a background paper for this report was commissioned. This is the first analysis for LAC that provides spatially explicit, quantitative estimates of historical GHG emissions resulting from deforestation activities (Harris et al. 2008). Results from this analysis provide information about the estimated magnitude of potential emissions in total for the Region, as well as identify specific countries and approximate locations within each country where efforts to prevent deforestation might result in the largest avoided emissions in the future. This high-resolution tool can effectively identify deforestation drivers and improve the targeting of policies and enforcement efforts by the institutions responsible for resource management and planning.

Notwithstanding the large variation in existing estimates, the available evidence suggests that the very large mitigation potential existing in this sector could be tapped at a relatively low cost and with significant synergies with other sustainable development objectives. In this regard, and considering that under a business-as-usual scenario future deforestation rates are estimated to remain high in South America and other tropical areas, it appears that mitigation activities in this sector should be a top priority for the Region (assuming there is adequate future international demand for this type of GHG mitigation efforts).

Transport

The LAC Region's transport sector is fast growing in terms of GHG emissions because of the rapid economic growth and the associated rise in car ownership and use, a modal shift away from public transportation to private vehicles, and the rising length and number of trips per vehicle as cities sprawl. With an average of around 90 vehicles per 1,000 people, the motorization rate in the LAC Region exceeds those of Africa, Asia, and the Middle East, even though it is still less than half of that in Eastern Europe and a fraction of the OECD countries' rate of nearly 500 vehicles per 1,000 people.[64] In Mexico—the second largest country in the region after Brazil in terms of the absolute level of transport sector emissions—car ownership is expected to increase at an annual rate of 5 percent from a fleet of 24 million in 2008 to 70 million vehicles in 2030.[65] Motorization rates are rising in the region in tandem with increasing incomes and improved availability of low-cost vehicles (box 1).

With the current growth in vehicle ownership and use, especially in urban areas, there is a pressing need to address issues related to emissions from private vehicles. In addition, traffic congestion in urban areas and a large share of highly polluting and inefficient vehicles on the road have meant that transport is also the leading cause of air pollution in Latin American cities. The rapidly rising emissions and large benefits from local environmental improvements mean that the transportation sector in the LAC Region offers significant potential for mitigation—especially when institutional barriers can be overcome—while at the same time delivering important auxiliary benefits.

Many no-regrets mitigation measures are available in the transport sector that can be implemented either with large savings or at a relatively low cost but with significant cobenefits. Time savings, improved fuel efficiency, and health benefits from better transportation systems can offset a substantial fraction of mitigation costs.[68] For example, studies have calculated that for Asian and Latin American countries, tens of

---

BOX 1

Demand for Private Vehicles Is Rapidly Rising in Latin America

A growing middle class has helped spur the demand for private vehicles. A study in 2005 of low-income families in four former "favelas" (shanty towns) in São Paulo found that 29 percent of families owned a car.[66] Over the years, efficiency improvements and competition have led to a slow decline in vehicle prices, with vehicles becoming more accessible to larger groups of people. There is increased competition from inexpensive vehicles from Asia, and the second-hand vehicle market is also growing. Vehicle sales in Latin America are breaking records and are expected to continue to post solid gains, buoyed by economic growth. Brazil and Mexico are the largest auto markets in Latin America, but Peru is the region's fastest-growing market. During the first three quarters of 2006, vehicle sales in Peru soared by 41 percent. The latest trends worldwide have vehicle manufacturers developing sturdy and inexpensive vehicles, specifically and successfully advertised to the middle- and lower-middle-income classes. For example, in São Paulo the fleet is growing at a rate of 7.5 percent per year, with almost 1,000 new cars bought in the city every day. This has accelerated the motorization rate in already congested cities and caused a rapid deterioration of the existing transport systems and infrastructure. The result has been deteriorating air quality, numerous traffic deaths and injuries, millions of hours of lost productivity, and increased fuel consumption and consequently rising GHG emissions. According to *Time Magazine*, São Paulo has the world's worst traffic jams.[67] In 2008, the accumulated congestion reached an average of more than 190 km during rush hours, and on May 9, 2008, the all-time record was set at 266 km, which meant that 30 percent of the monitored roads were congested.

thousands of premature deaths from air pollution could be avoided annually from moderate CO_2 mitigation strategies in the transport sector.[69] In Mexico, many no-regrets measures in the sector are expected to have significant cobenefits (box 2). Despite the low or negative economic costs of these options after accounting for their complementary benefits, most of these "low-hanging fruits" have not yet been "harvested." Indeed, institutional and regulatory obstacles impede the implementation of some options, and others require that costly monitoring systems are put in place.

The region's main challenge in terms of reducing GHG emissions from the transport sector is to decouple growth in emissions from rising incomes, despite the higher rates of vehicle ownership that accompany income growth. In dealing with the transportation of people, the top policy priority in the region is to slow down the rapidly rising rate of emissions from light vehicles by providing incentives for more efficient cars and for reduced car use. This can only be attained with integrated transport strategies that span across different transportation modes and are supported by efforts to reduce urban sprawl through better urban planning. In the transportation of goods, optimization of freight traffic through better logistics and improvements in fuel efficiency of heavy-duty vehicles are the top priority.

Renewable energy

Renewable energy, including large-scale hydropower, has the potential to reduce significantly the use of coal, petroleum products, and natural gas in power generation. Hydropower has traditionally supplied the majority of electricity in countries such as Brazil, Colombia, and Peru, but the share of hydropower has been falling in recent years as gas-powered and thermal generation has provided a significant share of new generation.

LAC has considerable potential for renewable energy generation. Wind conditions are excellent in many LAC countries—for example, with a wind power class equal to or higher than 4. The best wind resources are located in Mexico, Central America and the Caribbean, northern Colombia, and Patagonia (both Argentina and Chile).[70] High solar radiation levels of more than 5 kWh/m^2—which is high by

international standards—exist along South America's Pacific coast, in northeast Brazil, and in large parts of Mexico, Central America, and the Caribbean. Geothermal resources are also significant, as many countries in the region are located in volcanic areas. The potential of biomass is also well proven, with biofuels already accounting for about 6 percent of the energy consumed in the region's transport sector, dominated by ethanol production and consumption in Brazil. The region's largest potential in the area of renewable energy, however, lies in hydroelectricity. The region's total potential in this area was estimated to be about 687 GW, spread among Mexico and South and Central America.

Some wind projects are competitive with liquified natural gas (LNG), diesel, and high-cost hydroelectric projects, both in a scenario that assumes oil prices at US$60/bbl and in one in which prices reach US$100/bbl.[71] Moreover, in Brazil, Chile, Colombia, Ecuador, and Peru, medium- and large low-cost hydroelectric projects—with levelized generation costs (including investment, operation and maintenance costs) below US$37/MWh—are competitive with all thermoelectric alternatives in the two above mentioned scenarios for oil prices.[72] The only exceptions would be gas-fired plants in the cases of Peru—given the low domestic price of natural gas at US$2.1/MBTU—and Colombia for a scenario of low international oil and gas prices. This evidence is consistent with the findings of recent studies that identify a significant potential for reducing GHG emissions at negative costs through the implementation of hydropower projects in Chile and Brazil—respectively, by about 5 $MtCO_2e$ and 18 $MtCO_2e$ per year. An even larger potential has been identified in the case of Peru—about 59 $MtCO_2e$ per year—although in this case mitigation costs would be low but not negative—US$7.0 per tCO_2e.[73]

Similarly, in Central America hydropower projects with investment costs in the range of US$2,000/kW and average levelized costs of about US$59/MWh would also compete with LNG-fired, combined cycle gas turbine (CCGT) plants and diesel engines for both oil price scenarios. While in these countries hydroelectric plants would not be able to compete with coal-fired generation plants, carbon prices as low as

BOX 2

Cost-Benefit Analysis of Mitigation Measures in Mexico's Transport Sector

An analysis of transport mitigation options in Mexico demonstrates that there are numerous cobenefits of transport options, including financial, time savings, and local environmental improvement. While there is considerable uncertainty regarding the exact numbers, among the options that may provide the largest GHG reductions in Mexico are vehicle inspection and maintenance programs (including import restrictions on high-emitting vehicles), optimized transport planning (including public transport and freight), vehicle efficiency standards, and urban density policies (box figure). The economic benefits resulting from these interventions include the financial benefits compared to alternative means of transportation, time savings to individuals, for instance

by reducing congestion, and the local health benefits resulting from decreased local air pollution emissions (accruing to both commuters and to local inhabitants). This leads to negative costs for reducing GHG emissions for many of the interventions evaluated. (The environmental health benefits are not included in the costs in the figure.) As is typical of such studies, other important costs that are difficult to estimate are not quantified, such as the costs of implementing monitoring systems, overcoming information failures, or policy or regulatory changes. However, given that most of these interventions have already been implemented on some scale in Mexico, these costs are viewed by transport experts to be "surmountable."

Mitigation potential and benefits in Mexico's transportation sector—including the gains from efficiency and time savings but excluding environmental benefits and regulatory and monitoring costs

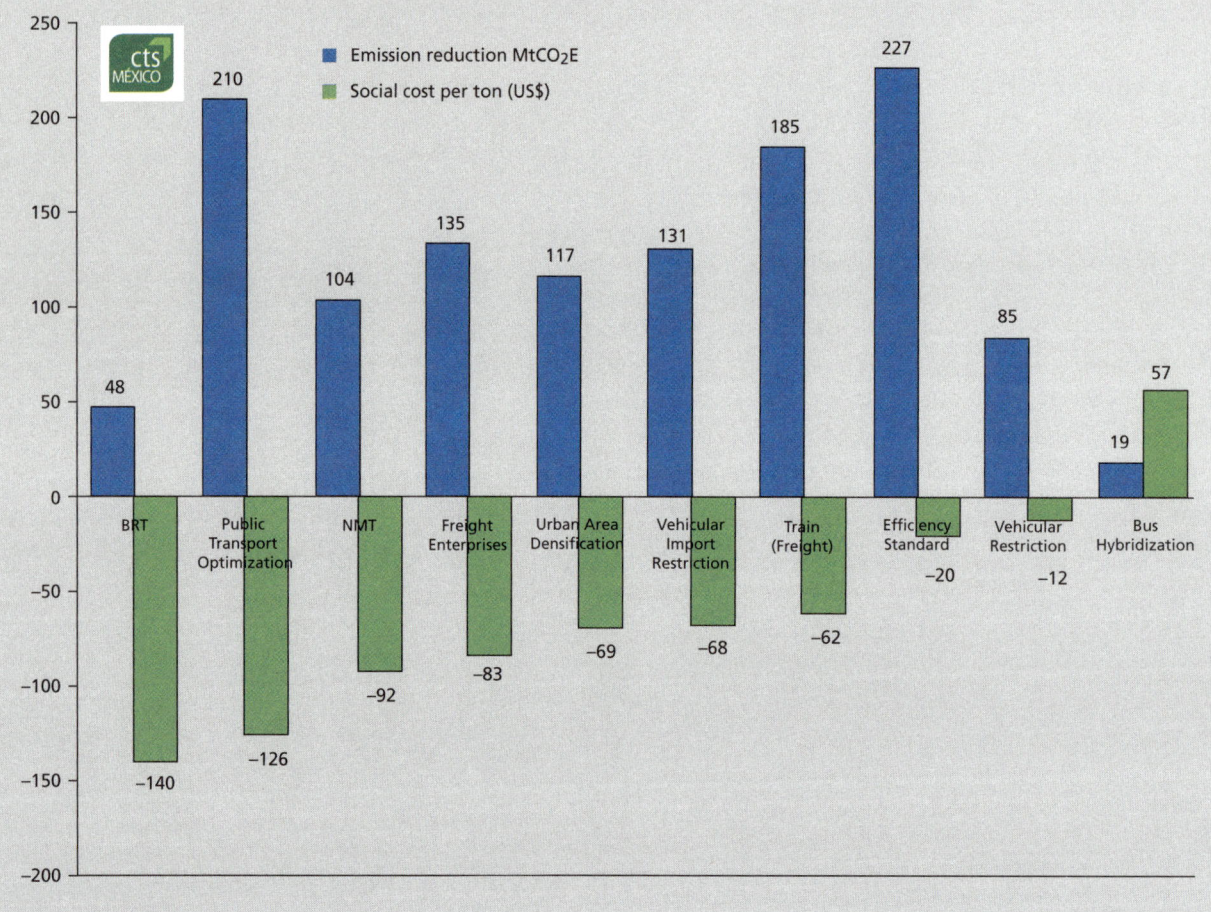

Source: MEDEC 2008.
Note: BRT is bus rapid transport. NMT is nonmotorized transport.

US$9/tCO$_2$ could equalize the costs of both types of alternatives, thus allowing a *switch* to the cleaner one at no additional cost. Much higher carbon prices would be needed, however, to make gas-fired plants competitive with their "dirtier" coal-fired counterparts—investors would have to assume carbon prices above US$25/tCO$_2$ to prefer the former over the latter. This suggests that if the opportunities for hydropower development and other renewables are not explored, several countries in the region—that is, those without access to low-cost natural gas—are likely to increase the carbon intensity of their fossil-fuel-based power generation capacity, thus leading to higher rates of GHG emissions.

Current expansion plans call for exploitation of only a small fraction of the region's hydropower potential—about 28 percent by 2015 (table 4), possibly rising to 36 percent by 2030, according to IEA projections. This is due in part to policy barriers existing in some countries: cheap fuel prices, cumbersome licensing processes, and unclear procedures for man-

aging environmental and social issues. Climate change impacts are creating another risk for hydroelectric plants, through accelerated glacier melt and variations in rainfall that need to be taken into account in planning and operating hydropower plants.

The effect of these challenges is illustrated by the case of Brazil, a country that has been very successful in developing low-cost hydroelectric generation, but has experienced delays in the development of new hydropower projects. Brazil has been using public auctions since 2004 to award long-term energy supply contracts. However, the participation of hydroelectricity in the auction process was constrained by delays in obtaining environmental licenses, and only about 50 percent of the hydropower projects that intended to participate in the first auction in late 2005 received an environmental license and were able to submit a proposal (World Bank 2008a). Consequently, the government decided to require that projects obtain at least preliminary environmental licenses before participating in auctions. Thus, the award of contracts for

TABLE 4

Largest Hydroelectric Potential in LCR (MW, % developed)

Country	Potential MW[a]	Installed 2004	Potential planned installed capacity by 2015	
			MW	%
Brazil	260,000	67,792	101,174	39
Colombia	93,085	8,893	9,725	10
Peru	61,832	3,032	3,628	6
Mexico	53,000	9,650	12,784	24
Venezuela, R. B. de	46,000	12,491	17,292	38
Argentina	44,500	9,783	11,319	25
Chile	25,165	4,278	5,605	22
Ecuador	23,467	1,734	3,535	15
Paraguay	12,516	7,410	9,465	76
Guyana	7,600	5	100	1
Costa Rica	6,411	1,296	1,422	22
Guatemala	5,000	627	1,400	28
Honduras	5,000	466	1,099	22
Panama	3,282	833	1,300	40
Total	646,858	128,290	179,846	28

Sources: a. Potential: OLADE estimates. SIEE Energy Statistics, 2006. Installed capacity by 2015 based on 2006 national expansion plans. EIA: Installed capacity 2004.

hydroelectricity in new generation capacity to be commissioned in 2008-10 has been lower than envisaged in the indicative generation expansion plans, and as a result the share of fossil fuel plants has increased. The government plans to facilitate investment in hydropower by conducting preinvestment studies and making them available to potential investors.

While motivated by legitimate concerns over environmental and social impacts, the environmental licensing process usually is lengthy, risky, and expensive. This can mean delays in the preparation and execution of the projects, and higher project risks and costs. The effect of such delays is hard to quantify, but one estimate is that a delay of one year in the commissioning of a hydropower project in Central America will increase the switching costs[74] from coal to hydropower by about 6.5 US$/tCO$_2$. Another recent study[75] estimated that in Brazil the cost of dealing with environmental and social issues in hydropower development represents about 12 percent of total project cost. Options for addressing some of these obstacles without compromising the environmental and social objectives of the licensing process are explored in section 5.

Notwithstanding the above-mentioned risks, there has been a renewed interest in the development of hydropower projects by both the public and, importantly, also by the private sector. Examples of the renewed activity include a substantial number of plants being built in Brazil, a recent auction in Colombia where the majority of winning projects were for hydropower, a plan to hold new auctions in Peru aimed at encouraging hydropower development, and the existence of small and medium-size entrepreneurs building hydropower plants in Honduras. Still, it must be recognized that the development of more than 100,000 MW of medium and large hydroelectric projects in South America and some Central American countries, included in the generation expansion plans by 2030, presents a considerable challenge.

As they do with other long-term investments—such as in hydropower—private developers of wind projects typically require long-term contracts with stable energy prices sufficient to recover their fixed costs. While wind power may be competitive today in

certain countries in comparison to fossil fuels, if oil prices fall in the future the opportunity cost may drop to levels that do not cover its costs. To address these hurdles, some countries have implemented quota-based incentive programs and long-term contracts with stable prices aimed at promoting the development of renewables. These and other policy measures to explore the Region's large potential in renewable energy are explored in more detail in section 5.

Renewable energy development offers substantial cobenefits. For example, decentralized electrification with renewable energy can provide large social and economic benefits to underserved populations that are usually dependent on traditional energy sources, such as biomass, kerosene, diesel generators, and car batteries. Compared to costly grid extensions, off-grid renewable electricity typically is the most cost-effective way of providing power to isolated rural populations. In Latin America, an estimated 50-65 million people still live without electricity. In Bolivia, Nicaragua, and Honduras, rural electrification rates are below 30 percent.[76]

Other potential cobenefits associated with increasing the share of renewable energy include the possibility of avoiding high-carbon technology lock-in, as discussed above, and providing some insulation from the high volatility of oil prices. With regard to this last point, LAC has a number of energy-importing countries that during recent years have been negatively impacted by increasing energy prices or decreasing fuel supplies.[77] The exposure to volatile oil prices is prompting countries everywhere to take measures to diversify their energy matrixes and to reduce the need for energy imports through increasing renewable energy generation and improving energy efficiency.

As for the risk of locking in technologies that could eventually become obsolete—given possible regulatory changes that would penalize emissions—it is worth noting that investments in long-lived capital assets in energy generation can last several decades. The Region is projecting a 4.8 percent annual rate of growth in electricity demand over the next 10 years, corresponding to a net increase of 100,000 MW in generation capacity, of which 60,000 MW are not under construction and have not been contracted.[78] The carbon intensity of this new generation capacity will be decided over the next few years as investment decisions are made. Policies and incentives that steer investment toward a low-carbon path will help the Region avoid installing technologies that in an increasingly carbon-constrained world will soon become obsolete, and make the Region lose competitiveness.

While the recent drop in oil prices makes renewable energy appear less competitive, a factor to be considered as part of the equation in evaluating renewable energy as an option for power generation is the volatility of oil prices, which increases the risks associated with thermal power generation costs (see box 3).

Biofuels

Liquid biofuels are one of a few existing alternatives to fossil fuels for transport. With oil prices reaching record highs during recent years, Brazil, the European Union, and the United States, among others, have actively supported the production of biofuels, based on various agricultural feedstocks—usually maize or sugarcane for ethanol and various oil crops for biodiesel. While the mitigation of climate change has been mentioned as one of the motivations for such support programs, there are other important objectives driving these programs. These include possible contributions to "energy security" and the possibility of rural employment generation and boosting farm incomes. Based on these supposed cobenefits, many governments in LAC and elsewhere are considering or beginning programs to encourage use and production of biofuels.

With few exceptions, development of biofuels poses several social and environmental risks. These include upward pressure on food prices, intensified competition for land and water, damage to ecosystems, and indirect impacts on emissions from land-use change—for example, when converting forests to agricultural production. These latter impacts are critical from the point of view of mitigation policies, as they could potentially eliminate biofuels' positive contributions. In summary, it has become increasingly clear that the costs and benefits of biofuels need to be carefully assessed before extending public support and subsidies to biofuels industries.

BOX 3

Incorporating Fuel Price Volatility in Power Planning and Investment

Generation of electricity with renewable energy, for example, using hydropower or wind, is characterized by local availability of the resource, high capital costs, and low and stable operating costs. These characteristics are different from those of thermal power plants, which are characterized by lower capital costs and higher operating costs, mainly for fuel. While future oil prices have always been uncertain, today's levels of price volatility are unprecedented, as demonstrated by the fall in prices in 2008 from US$150 per barrel to US$50 per barrel. This volatility increases the risk associated with the cost of electricity from a thermal power plant. Power system planners have traditionally tried to accommodate fuel price volatility by using different price levels of oil, gas and coal in their planning exercises. While these methods provide point estimates of the riskiness of a particular project or the sensitivity of a generation portfolio to the level of fuel prices, they do not address the issue of risk caused by price volatility. New techniques are being developed to take into account the value of a higher but stable cost option in comparison to a lower but more volatile cost option.

These techniques enable analysts to make specific tradeoffs between the return/cost of a generation option and its relative riskiness. This tradeoff between risk and return can also highlight the role of "free-fuel" renewables in the overall power generation mix. By combining the power of traditional generation expansion models with portfolio analysis techniques, it is possible to assess the relative risks and returns of a wide array of potential generation portfolios and to quantify the differences among them. Use of these methods permits the system planner or investment analyst to look at investment risks more systematically than has been the case in the past.

Brazil—the largest player in the global biofuels markets with about half of the global ethanol production—has developed the capacity to produce ethanol at a fraction of the cost of producing it in other countries. Because of favorable conditions for cultivation of sugarcane and the uniquely flexible industrial structure for sugarcane and ethanol processing, in periods of high oil prices Brazil's ethanol industry has been competitive even without government support. Brazil, in fact, may be the only country in which the ethanol industry has been able to stand on its own without government subsidy, and even in Brazil, this appears to have been the case only in 2004–05 (but not 2006 when international sugar prices skyrocketed) and 2007–08. (The Brazilian industry was also subsidized for many years to get to this point.[79]) Elsewhere, biofuels production has not been financially viable without government support and protection. Biofuels producers in the European Union and the United States receive additional support—over and above farm subsidies and support to producers through biofuels mandates and tax credits—through high import tariffs.

In evaluating the mitigation potential of biofuels, it is necessary to take into account the emissions coming directly from producing and burning them, relative to gasoline, and also emissions from land-use changes that come about from growing feedstocks. There are divergent assessments of the overall impact of biofuels on GHG emissions depending on which feedstocks are used to produce them and how those crops are grown. Without considering changes in land use, Brazilian ethanol from sugarcane may reduce GHG emissions by about 70–90 percent with respect to gasoline. For biodiesel, the emission reductions are estimated up to 50–60 percent with respect to gasoline. In contrast, the reduction of GHGs for ethanol from maize in the United States falls only in the range of 10 to 30 percent—also before taking into account the indirect GHG emissions from land-use change.[80] By some estimates, the cost of reducing one ton of carbon dioxide (CO_2) emissions through the production and use of maize-based ethanol could be as high as US$500 a ton.[81] The extent of the social risks—mainly the pressure that some biofuels put on food

prices—also varies by the type of biofuel. In contrast to large-scale diversion of corn for ethanol production in the United States, Brazil's ethanol production from sugarcane does not appear to have contributed appreciably to the recent increase in food commodity prices.[82]

Impacts on emissions from land-use change can arise directly, when feedstocks are grown in areas that were previously not used for agriculture, or indirectly when, for example, feedstock production displaces crop areas and pastures, which in turn expand into forest areas. The problem, however, is that when incentives are put in place to produce ethanol, it is impossible to assure that only low-productivity land will be converted, unless countries have in place adequate policies, institutions and transparent monitoring systems to safeguard other types of land from conversion. Even then, it is possible that the result may be land conversion in another country (see box 4).

LAC has the advantage of having large amounts of land devoted to low-productivity agriculture and pastures. To the extent that there is potential for increasing productivity in these areas, biofuels production could in principle increase without causing large increases in land use change emissions and while minimizing competition with food production. Whether this happens in practice would depend on how effectively land use change can be controlled. For countries considering whether and how to promote biofuels production, it is worth considering carefully whether the appropriate institutions and legal systems are in place to control land use change, and also whether the benefits outweigh the necessary fiscal and other costs.

Efforts are underway to develop sustainability certification schemes for biofuels, which in the long term could help reduce the environmental and social risks. The many obstacles to effective implementation of such schemes range from the need to ensure broad participation of all major producers to the difficulty, if not the impossibility, of accounting for indirect land-use change. For countries without the potential to produce low-cost first-generation biofuels, "second-generation" cellulosic technologies for producing ethanol from waste materials hold the promise of delivering GHG reduction benefits with lower social and environmental risks, but are still many years away from commercialization. In the meantime, it is clear that from the perspectives of emissions, social costs, and economic production costs, ethanol from sugar in Brazil is superior to alternatives. Reducing or eliminating the high trade barriers and huge subsidies currently in place in many countries would produce economic benefits for Brazil and its trade partners, and reduce GHG emissions.

Agriculture

The LAC Region has great mitigation potential in the agricultural sector, associated with the deployment of improved agronomic and livestock management practices, as well as with measures to enhance carbon storage in soils or vegetative cover. Some of these measures have significant cobenefits. Only about a third of this mitigation potential, however, could be economically exploited unless carbon were priced at

---

BOX 4

In Evaluating Biofuels' Impact on Overall Emissions, Land-Use Change Is Critical

The substitution of biofuels for petroleum-based fuels reduces emissions from vehicles to the degree that the former offset the GHGs released as they burn by sequestering carbon in their feedstocks. After appropriately accounting for this and other "life-cycle" effects (emissions involved in growing and processing feedstocks), emissions directly attributable to producing and burning ethanol from Brazilian sugarcane are estimated to reduce GHG emissions by 70 to 90 percent compared to gasoline. In contrast, the reduction of GHGs for ethanol from maize in the United States is only in the range of 10 to 30 percent.[83]

But the story does not end there. Land used to produce feedstock for biofuels—let's say maize—must be taken either from production of other crops or from some other current use. If the land for maize is converted from most other uses (forests, grasslands, pastures), GHGs are released as the soil is disturbed and as the vegetation removed from the land (which is sequestering carbon) is burned or decays. In evaluating the overall impacts of biofuels, this one-time release of GHGs is analogous to an up-front investment, which then must be "paid back" over time by the ongoing flow of emission reductions coming from the substitution of biofuels for gasoline.

If the land to grow more maize is taken from other crops, this in turn reduces the supply and raises the prices of those products. The higher price reduces consumption to some extent and also gives other producers an incentive to grow more. This increment in supply can come

from land being switched from yet other crops and/or nonagricultural land being converted. To the extent land is converted, it has the effect described above of releasing GHGs.

The original increase in maize production thus starts a chain reaction of land-use changes in the agricultural markets. Because global markets are well integrated, the original changes in the price of maize are transmitted globally, and so these indirect land-use changes may occur anywhere, not only in the country in which the biofuel feedstock takes place. An overall assessment of the impact of biofuels on GHG mitigation also needs to take into account the emissions resulting from both direct and indirect land-use change.

This type of indirect land-use change is particularly difficult to measure and because of that complexity it is often overlooked in sustainability assessments of biofuels. But the implications are enormous. For example, as noted above, life-cycle analysis indicates an annual saving of around 20 percent in CO_2 emissions relative to oil when ethanol is produced from maize in the United States. However, a recent study estimates that land conversion in the United States and elsewhere to produce more maize may actually result in a doubling of the GHG emissions over 30 years and increase GHGs for 167 years.[84] This study projected increases in cropland for all major temperate and sugar crops and livestock using a worldwide model as a result of an expected increase in U.S. corn-ethanol production by 56 billion liters by 2016.

(Box continues on next page)

BOX 4
(continued)

In that model, the resulting diversion of 12.8 million hectares of U.S. cropland would bring 10.8 million hectares of additional land into cultivation, of which 2.8 million are in Brazil, 2.3 million in China and India, and 2.2 million in the United States,[85] with the impact on GHG emissions depending on the type of land that is converted. Excluding indirect land-use change, Brazilian sugarcane is assumed to reduce emissions by 86 percent (with the carbon payback period of only four years) if sugarcane only converts tropical grazing land. An assess- ment in this study concurs with the conclusion from other studies that biofuels from waste have the most favorable carbon balance and questions the feasibility of reducing emissions through cultivation of dedicated feedstocks even on marginal land.[86] The findings regard- ing environmental costs of land-use change are corrobo- rated by studies that assess the carbon payback time for conversion of specific kinds of land, which indicate that ethanol from Brazilian sugarcane is clearly the most effi- cient in this regard[87,88] (see box figure).

Years needed to repay Biofuel Carbon Debt from Land Conversion (*)
(Ethanol from Corn or Sugarcane, Biodiesel from Soybean or Palm Oil)

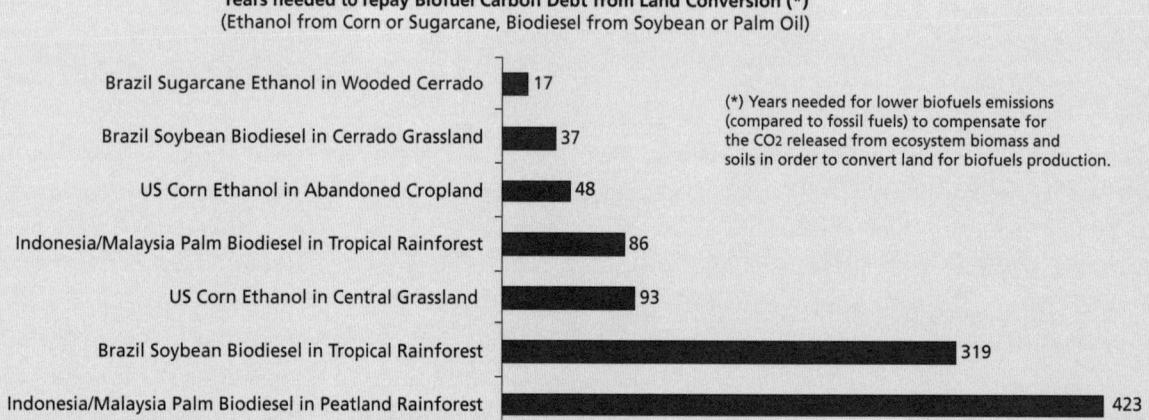

Brazil Sugarcane Ethanol in Wooded Cerrado — 17
Brazil Soybean Biodiesel in Cerrado Grassland — 37
US Corn Ethanol in Abandoned Cropland — 48
Indonesia/Malaysia Palm Biodiesel in Tropical Rainforest — 86
US Corn Ethanol in Central Grassland — 93
Brazil Soybean Biodiesel in Tropical Rainforest — 319
Indonesia/Malaysia Palm Biodiesel in Peatland Rainforest — 423

(*) Years needed for lower biofuels emissions (compared to fossil fuels) to compensate for the CO2 released from ecosystem biomass and soils in order to convert land for biofuels production.

Since the investment and the payback occur at differ- ent time periods, some argue that the payback flows need to be discounted, which might somewhat reduce the car- bon payback periods, but the choice of an appropriate discount rate for carbon is surrounded by political con- troversy and few studies have addressed this issue.[89] One recent study used a wide range of discount rates in an evaluation of this payback period with different kinds of land converted for ethanol in the United States and Brazil. It indicated a favorable cost-benefit analysis for some types of low-productive land in Brazil, using any of the discount rates considered.[90]

In assessing the impacts on overall emissions in pro- ducing biofuels in different countries, one relevant ques- tion is how much land must be shifted from other crops or converted to produce each gallon of biofuel. The ethanol yield per hectare from sugar in Brazil is about twice that of ethanol from corn in the United States.[91] This fact has led to the estimate that if the ethanol cur- rently produced in the United States were instead pro- duced in Brazil,[92] it would require only 6.4 million hectares, instead of 12.8 million, potentially leading to reduction in pressure for indirect land-use change and substantial savings in emissions from this source. But the potential for Brazilian sugar-based ethanol to replace less efficient production from other sources is limited by the current high barriers to import of ethanol into the United States and other high-income countries. Reduction of these trade barriers to imports of Brazilian ethanol could lead to substantial savings in world cost of production of ethanol and a lower level of land-use change.

over US\$20 per tCO$_2$e.[93] Obstacles to implementation that are specific to the agricultural sector include the issues of permanence of GHG reductions (particularly for carbon sinks), slow response of natural systems, and high transaction and monitoring costs.

Emissions from cropland can be reduced by improving crop varieties; extending crop rotation; and reducing reliance on nitrogen fertilizers by using rotation with legume crops or improving the precision and efficiency of fertilizer applications. In certain climatic and soil conditions, conservation or zero tillage can be effective both at improving crop yields, restoring degraded soils and enhancing carbon storage in soils. Methane emissions from ruminant livestock, such as cattle and sheep, as well as swine, are a major source of agricultural emissions in the LAC Region. Measures to reduce emissions from livestock involve a change in feeding practices, use of dietary additives, selective breeding, and managing livestock with the objective of increasing productivity and minimizing emissions per unit of animal products. Another approach in the case of animals confined in a relatively small area, like swine and dairy, is to use biodigestors to process waste and capture the methane for later use. This can either be flared (potentially generating carbon credits, since emissions from flaring are much less potent as GHGs than is methane) or used to generate electricity for on-farm or local use. Projects to do this are currently underway in Mexico and Uruguay.

The potential for cobenefits as well as the effectiveness and cost of mitigation measures from this palette of agricultural practices vary by climatic zone and socioeconomic conditions. Conservation or zero tillage—an agricultural practice that has been successfully applied over nearly 45 percent of cropland in Brazil—is a case in point. In contrast to conventional tillage, zero tillage involves no plowing of soils and incorporates the use of rotations with crop cover varieties and mulching (application of crop residues). The result is an increase in the storage (sequestration) of carbon in soils. Lower fuel requirements for plowing operations that are no longer needed are another source of GHG reductions. However, application of nitrogen fertilizers to counteract nitrogen depletion that often occurs in the first few years after conversion

from conventional to zero tillage may negate some of the reductions in GHG emissions.[94]

In summary, while there are a number of opportunities for contributing to increasing agricultural production while reducing GHG emissions, the proposed practices need to be evaluated within specific regional and local settings, and there is no universally acceptable list of preferred interventions. Furthermore, competition for land among different uses means that many solutions are more cost efficient and more effective at achieving reductions when they are implemented as part of an integrated strategy that spans agricultural subsectors and forestry. Since mitigation solutions are very context-specific in the agricultural sector, research efforts need to have a strong participatory dimension so as to ensure that they respond to the specific needs of small farmers.

Waste

The overall potential for GHG emission reduction through sanitary landfills and composting is not very large because of the low contribution of waste to LAC's overall emissions. However, proper collection and disposal of solid waste have very significant environmental, health, and public safety benefits, making this an important overall priority.

Inadequate waste collection and the resulting clandestine dumping of waste in cities increase the risk of flooding when waste blocks urban waterways and drainage channels; burning of waste on city streets or in open dumps emits carcinogenic dioxins and furans because of incomplete combustion and other contaminants; garbage dumps are a major source of leachates to surface and groundwater and they proliferate the spread of vector-borne diseases by insects, rodents, and birds. Solid waste disposal sites that do not have gas management systems accompanied by flaring or energy recovery are major sources of methane discharges, and leaking methane gas can explode in people's houses or in public areas.

Municipal waste collection rates are generally acceptable in LAC, particularly in larger cities in the region. On average, cities with more than 500,000 inhabitants collect over 80 percent of their waste. In smaller cities, however, technical and financial difficulties result in a lower collection rate of around 69 percent. Overall, 62 percent of the waste generated in LAC is burned or ends up in unknown disposal sites.[95] The good news is that solid waste management is high on the political agenda of local governments and many mitigation measures that also have large local cobenefits can be implemented at modest incremental costs. In fact, many examples of successful implementation of waste management strategies can be found in Mexico, Brazil, and Colombia, among other LAC countries. Emulating these examples of good practices could have an important positive impact.

5. Policies for a High-Growth, Low-Carbon Future

Keeping the countries of LAC on a trajectory of high growth and poverty reduction, while at the same time maximizing their contribution to reducing global emissions, will require a coherent set of policies on three levels. First, given that climate change is inevitable—indeed it is already happening—the countries of the Region will have to adapt their own growth and poverty reduction strategies so as to minimize the adverse impacts on their populations and ecosystems. Second, in order for global mitigation efforts to be effective, efficient, and consistent with equity considerations, there must be an appropriate international policy environment in place, including (a) full participation by the high-income countries in an agreement on climate change and (b) a LAC-friendly global climate change policy architecture. LAC countries can actively take a leading role in the negotiation of this agreement and the implementing architecture. And third, in order for the LAC countries to exploit the various efficient mitigation opportunities described in the previous section, a series of new domestic policies will be required.

Adapting efficiently to a changing climate in LAC

Introduction

Just as they have adapted to past climatic shifts, humans and ecosystems will, to some extent, spontaneously respond to the forthcoming climatic changes in ways that will reduce the negative effects and enhance the positive. In this context, a major challenge for governments and the international community will be to provide the policies, institutional infrastructure, and public goods that will facilitate and support the autonomous process of adaptation of human and natural systems. One-size-fits-all strategies, however, will not work well in dealing with climate change, as the way in which individuals adapt will be highly idiosyncratic. Moreover, to the extent that most individual adaptive actions will have little effect on others—that is, they will involve small or no externalities—most government policies to support human adaptation will probably have to be "facilitative" in nature (Tol 2005). In other words, governments may need to focus on nonprescriptive measures that establish a framework for individuals to adjust, and empower them to do so, but do not direct them how to change behavior, nor subsidize private investments. The main objective should be to expand options and enhance households' economic resilience and mobility—their ability to make well-informed decisions and welfare-enhancing economic transitions in the face of longer-term changes in the external environment.

Not all adaptation policies, however, will be facilitative. There will of course be areas in which governmental interventions and investments are necessary to deal with climate change, just as they now deal with natural disasters—both to help prevent damage and to aid in recovery. Active interventions by governments and international institutions will be necessary to provide some critical public goods, including improvements in natural resource management systems, infrastructure investments to provide direct protections against climate-related threats, and additional investments in the development and deployment of technologies that will be critical for producers to adapt to climatic changes. Beyond the provision of these public goods, facilitative policy responses will be important in the areas of weather monitoring and forecasting, social protection, climate-related risk management, and improvement of water and financial markets. In most of these cases, we argue, adaptive responses will be highly congruent with good development policies. In other words, mainstreaming climate change considerations in government policies will often involve measures of a no-regrets nature.

Necessary public policy actions to adapt to CC that go beyond facilitation

The nature of climate change itself and several inherent features of adaptive responses will be relevant in shaping optimal government policy. As we have seen, climate change is both long term and in important respects uncertain in its effects on weather in specific locations. Undertaking major investments or policy responses in anticipation of specific future climatic impacts runs a high risk of wasting resources or even increasing adverse impacts if the changes do not materialize as expected, or if future technological advances allow a more cost-effective response. Weighed against that is the risk that failure to take timely actions may incur preventable damages, and some investments and policies may take a long time to bear fruit. The need to strike a balance between these considerations argues that policy should be flexible over time, easily allowing updating as new information becomes available—for example, investments in coastal protection that allow for expansions as new information on the risk of sea level rise becomes available. There is value in waiting for more information and better technology, so nonurgent decisions may be deferred, and investments should be designed in modular ways when feasible. This said, some of the main areas in which public policies will be critical to make adaptation to climate change both effective and efficient include the following.

Strengthening natural resource management, focusing especially on managing changing water flows and improving resilience of ecosystems. In addition to providing a supportive environment for development of water markets, governments may need to invest directly in public goods to improve drainage in areas with increased rainfall or in new dams to regulate the flow of water in areas where glaciers have melted and no longer perform this function. On the other hand, some dams may need to be decommissioned as they may no longer be needed if flows fall sufficiently. This

is one area in which the mitigation and adaptation agendas may intersect, in countries where multi-use dams could help manage flood control while also generating clean electricity.

Public investments will also be needed to preserve ecosystem services in the face of climate change impacts. One key short-run component in a strategy to help ecosystems adapt to climate change over the next few decades will involve reducing other stresses on those systems and optimizing their resilience. In the next decades, as conditions change and more information becomes available, other potential strategies can be identified. Biological reserves and ecological corridors can serve as adaptation measures to help increase resilience of ecosystems (Magrin et al. 2007). Helping coral reefs survive in an environment of rising sea surface temperatures, for example, may require increased attention in the design of marine protected areas to identifying and protecting particular reefs that are especially resilient, either because they are located where cool upwelling provides natural protection against thermal events or because they seem to have natural resiliency.[96] Some ecosystems or individual species may need to be "transplanted" to more hospitable environments as their current habitats become too hot, or at least corridors preserved so they are able to migrate. Recent projects to preserve the coral reefs in the Caribbean and protect the integrity of the Meso-American Biological Corridor are examples of this kind of effort, which can be scaled up in the future.

Investment decisions in activities to support ecosystem adaptation must be based on sound science, underscoring the need to build capacity in the Region and the need for transfer of resources for this purpose. The foundation of more reliable vulnerability and impact assessments is the availability and use of sound science. Resources for strengthening the capacity of the local scientific community and relevant governmental institutions in LAC, and transfer/sharing of knowledge from the developed world are necessary for the development of an adaptation agenda. This is the focus of a number of ongoing projects in the region (box 5).

BOX 5

Climate Change Projects in LAC

Current projects in a number of countries focus on building capacity and generating knowledge to assess vulnerabilities and risks associated with climate change, particularly those related to ecosystems. Some examples of these activities, which are being carried out in partnership with local academic and research institutions, include:

- Expansion of the coral reef monitoring network through the installation of a coral reef early warning station (CREWS) in Jamaica and the update of sea level monitoring stations in 11 countries in the Caribbean.
- Generation of climate projection scenarios in the Caribbean focused on adapting existing global climate change models to develop appropriate statistically and dynamically downscaled regional climate change models relevant for the region. The

results of this effort have served as input in the preparation of national adaptation strategies.

- Application of data from the Earth Simulator of the Meteorological Research Institute of Japan (MRI) for the design of basin vulnerability maps in the tropical Andes (Bolivia, Ecuador, and Peru). This effort is being complemented with the installation of a monitoring network of eight high mountain meteorological stations to measure the gradual process of glacier retreat, and development of a climate monitoring system to analyze the carbon and water cycle in "*paramo*" ecosystems in the Tropical Andes.
- Development of a methodology for the assessment of impacts of anticipated intensified hurricanes on coastal wetlands and quantification of these impacts in Mexico.

Strengthening direct protection against climate-related threats in cases for which collective action is needed. Some investments have characteristics of public goods in that the benefits are shared by all and individual payments would be infeasible to organize. These would include investments to "climate-proof" public infrastructure, control floods, better regulate more erratic water flows, and protect coastal populations in the face of rising sea levels. Many of these will need to be carried out at local levels of government. For example, more intense rainfall will threaten to overwhelm sewer systems in cities where storm sewers are not separated from sanitary sewers, requiring that these systems be rebuilt to avoid threats to public health. Measures will be necessary to combat public health threats from vector-borne diseases as well. In connection with the latter, surveillance and monitoring will be especially important in those countries where it is expected that climate change will allow the expansion of disease vectors into new areas where the population lacks immunity. One project now underway, for example, focuses on strengthening of the public health surveillance and control system in several Colombian municipalities based on climate change considerations. The pilot program is setting up an early warn-

ing system based on the incorporation of system tools in public health surveillance to detect increases in the transmission of malaria and dengue, and aid in development of preventive strategies.

Where the effects of ongoing climate change are already being felt (for example, glacier melt in the Andes), infrastructure investments may be needed in the near future. A first step is now being taken with a project to help assess the impact of climate change on the hydrology of specific basins in Peru and the threat that this presents to water availability for drinking, agriculture, and generation of hydropower. For longer-term planning, the possibility of future climate change needs to be taken into account in a number of ways. Increased intensity of hurricanes—and possible increased frequency—implies that risks need to be re-evaluated, which will in turn mean that more climate-resistant engineering designs will pass the cost-benefit test. This is already being recognized in projects to help Caribbean countries recover from recent hurricanes, as infrastructure is being rebuilt to higher specifications.

But of course this does not necessarily mean that all investments to help harden infrastructure against anticipated climate change need to be started immediately. In conditions of uncertainty, when some of the uncertainty will be resolved as time passes, there is value in waiting, and this should be incorporated in planning. Tools for cost-benefit analysis that explicitly take into account this kind of uncertainty—such as real options analysis—will be useful in this regard. This will mean postponing actions in some cases and in others will lead to building in more flexibility by, for example, modular design of infrastructure.

Strengthening technological linkages and knowledge flows. Adoption of improved technologies could potentially minimize the kinds of adverse impacts on agricultural productivity that were quantified in section 2. Farmers in temperate regions should be able to adapt to warmer temperatures using existing varieties that are currently grown in more tropical zones. That is, varieties grown in warmer climates can be transplanted to warming environments, moving from low to high latitudes. This assumes that trade and regulatory regimes are open to such technology transfer. One issue that governments may need to consider is whether their regulations governing introduction of new varieties (GMOs and non-GMOs) should be revised in light of the increased value of technological "spill-ins" from abroad.[97] The cost-benefit calculus on which these regulations are based could be profoundly affected by climate change.

To the extent that existing varieties can in general satisfy the needs of farmers in areas that are not at the extreme ranges of crop tolerances, these conditions may not need to be the major focus of research and development of new varieties. In such cases, research may need to focus on the productivity limitations for crops that are currently being grown in areas close to their thresholds of temperature tolerance. This, however, may be a challenging endeavor. Many crops in LAC are grown in very thin temperature and rainfall ranges and may be susceptible to these threshold effects (Baez and Mason 2008). The problem is illustrated by the experience of The Brazilian Corporation for Research in Agriculture (Embrapa) in developing genetic varieties of crops that are more tolerant to high temperatures and water deficit, as well as to diseases and pests (cassava and banana hybrids). Embrapa has discovered that biotechnology can help crops deal

with climate stresses and increases in temperatures up to 2°C. Above that temperature, the efficiency of genetic improvements will be limited as it will hinder photosynthesis (Assad and Silveira Pinto 2008). And in any event, technological improvements take time to materialize and are costly. It takes between 5 and 10 years for new varieties to be developed and released, and perhaps even longer for them to be adapted to specific agro-ecological conditions.

Facilitative adaptation policies

The point is often made that good development policy is good adaptation policy. Higher incomes and human capital increase resilience to shocks of all kinds and give households the capacity to deal better with change. This point is well illustrated by a kind of natural experiment in Mexico's Yucatan Peninsula, where two hurricanes hit the peninsula 22 years apart. Hurricane Janet hit in 1955 as a Category 5 storm and killed over 600 people. Hurricane Dean landed in almost the same spot in 2007 as a slightly stronger storm, but with no loss of life. In the intervening 22 years, of course, private incomes had increased and government institutions had developed, allowing everyone to be better prepared.[98]

The fact that adaptation policy and development policy have much in common is good news in that the tradeoffs in deciding whether to take actions now or postpone them are not as stark. For many measures that are good economic policy, but may face political opposition or are currently low priority, the specter of climate change may alter the political calculus in a reform-friendly direction. For these, there is no reason to delay action. And there are other areas in which urgent action is warranted to deal with ongoing climate change or to prevent irreversible damages, especially to ecosystems that are currently under climate-related stress. For other measures, however, the high levels of uncertainty associated with predicting long-term changes in climate create risks that may outweigh any advantages of quick action. What is needed is a kind of triage or prioritization of actions to identify what has to be done in the short term and what can be postponed. The following are some of the most important examples of policies that facilitate adaptive responses and are in general good development policy.

Strengthening weather monitoring and forecasting tools

This will provide better information to reduce uncertainty and help people make well-informed choices. Some of the types of tools most valuable to reduce uncertainty are an historical climate database, weather-monitoring instruments, systems for analyzing climate data to determine patterns of intra-annual and interseasonal variability and extremes, and data on system vulnerability and adaptation effectiveness (for example, resilience, critical thresholds) (FAO 2007). For example, recent studies have quantified the potential economic value of climate forecasts based on predictions of the "El Niño-Southern Oscillation" phenomenon (ENSO[99]). They have concluded that increases in net return from better forecasting and consequent adjustments in agricultural production practices could reach 10 percent in potato and winter cereals in Chile; 6 percent in maize and 5 percent in soybeans in Argentina; and between 20 and 30 percent in maize in Mexico, when crop management practices are optimized (for example, planting date, fertilization, irrigation, crop varieties). Adjusting crop mix could produce potential benefits close to 9 percent in Argentina. (IPCC 2007, Ch. 13). The provision of reliable forecasts jointly with agronomic research has led to a drop in the damage of crops in drought times in areas of Peru and Brazil (Charvériat 2000). Yet in LAC, even the hardware is inadequate and in some cases the situation has become worse over time as weather data collection infrastructure has deteriorated. The density of weather stations has been diminishing for most countries in the Region, in part because of fiscal constraints in the maintenance of equipment and trained personnel. In Bolivia, for example, there are currently around 300 working weather stations out of 1,000 stations a few years ago. Likewise, Jamaica is currently operating around 200 weather stations, down from a total of 400 in 2004, and similar situations can be found in Guatemala and Honduras. Putting in place effective mechanisms for disseminating weather information is

also critical. Consultations in LAC countries have shown that even where weather information is in principle available, it is not well disseminated to stakeholders.

Strengthening social protection

Evidence reveals that food and basic nonfood consumption, education, health, and nutrition are particularly vulnerable to shocks. Well-targeted, scalable, and countercyclical safety nets can help keep the poor from falling into a "permanent poverty trap" and being forced into "low-risk, low-reward" production strategies or liquidation of productive assets in response to a weather shock. Several countries in the LAC region have been in the forefront of developing the conditional cash transfer as a safety net tool, with programs such as *Familias en Accion* (Colombia), *Bolsa Familia* (Brazil), *Red Solidaria* (El Salvador), *Oportunidades* (Mexico), *Red de Proteccion Social* (Nicaragua), *Programa de Asignacion Familiar* (Honduras), and *Atencion a Crisis Pilot,* a pilot program in Nicaragua specifically designed to respond to weather shocks.

There is considerable evidence that these programs can be effective in response to shocks of various kinds. Rural households in the area of influence of the *Oportunidades* program in Mexico have constant interactions with natural hazards: based on six rounds of surveys between 1998 and 2000, around 25 percent of them experienced a natural disaster. After such shocks, many families are forced to remove children from school, risking descent into a multigenerational poverty trap. But the indirect insurance offered by the program results in one additional child staying in school for every five children protected (de Janvry et al. 2006). And in response to the coffee crisis in 2000–03, the consumption of participants in the *Red de Proteccion Social* program in Nicaragua fell by only 2 percent, compared to over 30 percent for non-participants (Vakis et al. 2004). Similar results have been found for the *Programa de Asignacion Familiar* in Honduras to protect the consumption and investments in child human capital of coffee-growing households enrolled in the program in the face of the coffee crisis (World Bank 2005a). Social funds have also proven to be a good instrument to increase resilience to climate shocks and have the advantage that they can respond rapidly (Vakis 2006) (box 6).

Of course, each type of safety net has its strengths, flaws, and implementation challenges, and their effectiveness is likely to vary across countries and weather shocks. No one size fits all when it comes to design of effective interventions, and the choices of policy makers need to account for this degree of heterogeneity among different programs. Some specific features may need to be incorporated to tailor these instruments to weather shocks; for example, conditionalities to discourage exposure to climate risk.

The novelty of the *Atencion a Crisis Pilot* in Nicaragua—which was specifically designed with weather risks in mind—was to add two interventions (vocational training and a productive investment package) to the standard nutrition and education package to improve the resilience of poor rural households to natural risks and economic downturns.

In particular, these interventions intended to reduce the use of inefficient and costly (in terms of human welfare) ex ante risk management and coping strategies. Indeed, evaluation has shown—in addition to the effects on consumption, education, and nutrition—that these supplementary packages improved income diversification and the use of savings ex ante and reduced the use of child labor and the sale of assets to cope with shocks. Other lessons for program design are that it is important that the program be designed to scale up and down quickly, and that payments be well targeted. Two approaches to targeting are (a) preshock eligibility based on degrees of risk exposure and poverty/vulnerability, and (b) ex post targeting that incorporates actual levels of damage and impacts.

Strengthening households' and governments' abilities to manage risks, especially weather shocks
In order to facilitate private adaptation efforts, it is important to strengthen private insurance markets,

---

BOX 6

Social Funds and Natural Disasters: The Example of the Honduras Social Investment Fund and Hurricane Mitch

Despite the fact that Hurricane Mitch killed thousands of Hondurans, left a million homeless, and inflicted damage equivalent to two-thirds of GDP, poverty rose only moderately in its wake.

This remarkable reality is attributable largely to the efficacy of the Honduras Social Investment Fund (FHIS), a public program created in 1990 to finance small-scale investments in poor communities. Originally conceived as an antidote to the adverse effects of structural adjustment policies, FHIS nimbly became an emergency-response program of sorts after Mitch devastated the country in 1998.

FHIS successfully prevented the disaster from aggravating poverty by rejuvenating economic activity, and restoring basic social services. Within 100 days of the hurricane, the program approved US$40 million for 2,100 community projects; by the end of 1999, FHIS had financed 3,400 projects, four times the number financed in a comparable pre-hurricane period. Projects

prioritized clearing debris and repairing or rebuilding water lines, sanitation systems, roads, bridges, health centers, and schools, thus hastening national recovery and generating about 100,000 person-months of employment in the three months following the crisis.

The decentralized structure and institutional flexibility of the FHIS enabled its rapid and influential response. Building on strong pre-existing partnerships with municipalities and communities, FHIS directors established 11 temporary regional offices and quickly delegated resources and responsibilities. Directors reduced the number of steps in the subproject cycle from 50 to 8, established safeguards to ensure accountability and transparency, and effectively accessed International Development Association financing. As an article reviewing program outcomes concluded several years later, "FHIS demonstrates that a social fund can play a vital role as part of the social safety net in times of natural disaster."

particularly to address specific weather shocks. Among developing regions, LAC is second only to Asia in premiums for weather insurance, but the market is still very small. Furthermore, index-based weather insurance, which is probably in the long run the most viable form, is still a relatively foreign concept in most countries, notwithstanding significant technical assistance to introduce it. To grow this market, a number of obstacles need to be resolved. One is that insurance markets as a whole are underdeveloped in LAC. Measured by premiums as percent of GDP, LAC lags the developing regions of Asia, Africa, and Eastern Europe (Swiss Re). Another is the lack of a regulatory framework conducive to this type of insurance in most LAC countries. A third is that local insurers are unable or unwilling to take on the risk associated with catastrophes. One lesson of experience in providing technical assistance to develop this market is that sometimes governments may need to take this high-risk market segment, perhaps laying off some of the risk in international reinsurance markets. The vacuum in weather data is also a problem, and as noted above, this seems to be getting worse. International institutional innovations such as the Caribbean Catastrophic Risk Facility are helping governments in this Region manage their own risk exposure, and work is underway to develop a similar facility for Central America. But it has to be recognized that while insurance can help cope with short-term weather shocks—which may become more severe in the future—it cannot compensate for long-term climatic trends. And governments may need to adjust their own internal insurance policies—and their policies of damage compensation. If these insure people against their own risky behavior by compensating them for losses from weather risks, such policies can undermine incentives to adapt appropriately to changing climate.

Strengthening markets

On a national level, two kinds of markets deserve particular priority because they are currently poorly developed in most developing countries and because they will be especially important in making adjustments to climate change.

1. *Water markets.* Many of the most important impacts of climate change will be intermediated through water availability, yet water rights are currently ill-defined and water grossly undervalued in most countries. In virtually every water system around the world,[100] extensive amounts of water are currently used to grow low-value crops. In LAC, Chile and Mexico have made considerable advances, yet even in these countries, the markets are far from being adequately designed to allocate water to its highest valued use. Studies indicate that shifting water to its most valuable use can significantly reduce the harmful effects of climate change. One background study for this report used a simple illustrative simulation exercise to quantify the economic cost of water shortages forecast for the Rio Bravo basin in Mexico by 2100.[101] In one "maladaptation" scenario, the shortage was accommodated by across-the-board proportional reductions in all types of uses (agriculture, industry, and residential). In another scenario, the water was allocated to the highest-value uses, as would occur if it were efficiently priced. The economic costs under the former scenario were hundreds of times their size under the latter, underscoring the ability of efficient adaptation policy to reduce the costs of climate change, while not foreclosing complementary measures to address adjustment costs and distributional implications. In some cases, transbasin transfers may be useful in dealing with regional scarcity, as they have been in California. In LAC, potential for this kind of option exists in the Yacambu basin (República Bolivariana de Venezuela), Catamayo-Chira basins (Ecuador and Peru), Alto Piura and Mantaro basins (Peru), and São Francisco basin (Brazil) (Magrin et al.). But organizing such transfers will require considerable planning, investments, and in some cases international coordination. Effective international institutions will be necessary not only to facilitate transboundary water trade, but also to improve mechanisms for

mediating conflicts provoked by changes in water availability (UN Foundation).

2. *Financial markets.* Financial markets play two roles with respect to adapting to climate change. In the short term, they allow individuals to adjust efficiently to shocks through saving and dissaving to smooth consumption. In the longer term, financial institutions are sources of investment capital that will be needed to finance adaptation expenses. While urban areas in many LAC countries are reasonably well served by financial institutions, rural areas—especially small farmers—are generally not, for reasons related to high transactions costs and low ability of such clients to offer reliable collateral. Yet there are good examples of how these barriers can be overcome. Social capital and peer monitoring can be used to good advantage. Using a value-chain approach, for example, FUNDEA in Guatemala finances inputs and outputs for small farmers, accepting standing crops as collateral. Furthermore, public policy can support pilot testing of technological innovations that reduce costs and risks of offering financial instruments to rural small-scale producers. Just as cellular phones can speed market and price information to producers, so-called mobile or m-banking, now being piloted in Brazil, can also dramatically reduce transactions costs for rural financial transactions.[102] Where necessary, financial regulations may need to be reformed to remove interest rate ceilings and permit institutions to mobilize savings deposits, perhaps via branchless banking, taking advantage of existing post offices, gas stations, and other retail outlets as conduits for rural financial transactions. Stimulating data collection via credit-reporting bureaus can also reduce the current risk premium associated with rural lending, owing to information deficits to gauge behavioral risk of potential borrowers. Rural finance for smallholders could also benefit from the creation and expansion of insurance instruments to protect against losses, and in some countries, insurance has been packaged with microcredit.

In connection with the consumption-smoothing role of credit markets, the nature of weather-related shocks has an important policy implication. Weather shocks tend to be highly correlated across fairly large areas. This means that a financial institution with a client base concentrated in one area—particularly a rural area, where many clients rely directly or indirectly on agriculture—is likely to be poorly equipped to deal with a shock, since all of its depositors would need to withdraw savings at the same time. One way to deal with this is to insure the loans against weather risk. The other strategy is to rely on geographic diversification. Regulatory policy can encourage reliance on insurance by, for example, putting a premium on insured loans when calculating capital adequacy ratios. Alternatively (or in addition), it can promote the development of financial institutions with clientele that are not exclusively rural, and that are not heavily exposed to weather risks. In small countries especially, foreign banks may be best placed to fill this role, but in any case, regulatory policy could be designed to encourage development of extensive linkages outside of a rural client base.

A critical mass of participation by high-income countries is essential

Especially in the area of mitigation policies, strong leadership by all rich countries is a precondition for progress in the fight against global warming, for example, through a global agreement to which all these countries are signatories. This is important not only to set an example for other countries moving to a low-carbon growth path, but also to create the perception that such an agreement is equitable, thereby lending it credibility. From an economic perspective, this kind of participation is also necessary to create a market of sufficient size to generate incentives for the investments in research, development, and production that would be required in such a large-scale undertaking. The market could to a large extent be driven by the incentives created by valuing carbon emissions, whether through some kind of carbon tax or an international cap-and-trade system. Individual countries are likely to also have local regulations, taxes, and subsidies of various kinds. To the extent practicable, however, the system as a whole would ideally generate a net price of carbon emissions that is uniform across countries and activities.

Apart from agreement to take aggressive actions to reduce their own emissions, action by the high-income countries is needed in several other areas, as described below.

The need for high-income countries' leadership in technology development and transfer

While the pricing of carbon will automatically create incentives for progress in technologies for emission reduction, the public-good nature of knowledge will require public funding of some kinds of research, to support both mitigation and adaptation to climate change in developing countries. This is the case for basic research (to generate knowledge that has no short-term commercial application) and especially for research dealing with technologies the primary market of which is in countries where the population has low purchasing power. High-income countries have the skills and commercial base to undertake research and development of cutting-edge technologies for low-carbon power generation and energy efficiency. Much of the low-wind-speed technology now being employed in wind farms in the region, for example, is German, while technology to modernize bus fleets with hybrid engines comes from Japan, Brazil, and the United States. Some of this technology uptake has been financed through carbon finance (CDM), and small-scale donor projects have for years financed investments in clean technology such as microhydropower plants in Peru and solar powered irrigation pumps in Brazil. But more innovative ways need to be found to accelerate this process in the future. Various ideas have been advanced on mechanisms through which donors could encourage development and diffusion of technology in such countries. Mechanisms could include advanced commitments to purchase some set quantity of goods, purchasing existing intellectual property rights to make the technology widely available, or offering prizes for specific types of technologies.

Support for international research on climate change itself will be important, as will research on adaptation. Particularly important will be technologies to maintain agricultural productivity. In this sphere, private seed companies are investing significantly in developing varieties, including GMOs, with characteristics needed to cope with changing climatic conditions. But they cannot be expected to focus on open-pollinated varieties that would be most useful for small-scale producers in developing countries. For this, internationally supported research through the CGIAR (Consultative Group for International Agricultural Research) centers will be required.

Financing of human and ecosystem adaptation in developing countries

As discussed in section 3, equity considerations call for high-income countries—which bear primary responsibility for the GHGs that are causing global warming—to subsidize the consequent adaptation costs in developing countries, perhaps taking into account the varying degrees of responsibility and capability of different countries. The mechanism through which subsidies are administered is important, and should ideally be consistent with the economic principles that will shape adaptive behavior. Since adaptation policy largely coincides with development policy, it may make more sense to simply augment aid flows through existing mechanisms (multilateral and/or bilateral), rather than creating new mechanisms, provided that (a) this funding is transparently additional to normal flows and (b) aid is concessionary, even to middle-income countries.

In addition to supporting human adaptation to climate change, it is incumbent on high-income countries to provide financial and technical support for developing countries to preserve the global public good of biodiversity. Many LAC ecosystems threatened by climate change are of global significance. Internationally funded adaptation projects are already being piloted through the Global Environmental Facility (GEF), and successful ones can be scaled up and replicated. There is also an adaptation component in the new Climate Investment Funds managed by the World Bank, to which donor countries can contribute.

Maintaining an open international trade regime to facilitate efficient adaptation and mitigation

While all the countries that are members in the World Trade Organization (WTO) will play a role, leadership by the high-income countries will be critical in reaching agreement on some of the issues in the WTO that are particularly relevant for helping the world deal with challenges created by climate change. First, all kinds of barriers to food trade will need to be effectively disciplined. This would facilitate changing patterns of food trade as climate change alters production patterns over the long term, as well as spread the effects of short-term supply shocks and ensure that consumers and producers respond appropriately. With a share of close to 11 percent of world agriculture and food exports, LAC is currently a major food-exporting region. But some countries may suffer large losses in productivity, leading to dramatic shifts in food trade patterns inside and outside the region. This issue is therefore of vital concern to the LAC Region.

One of the lessons of the recent precipitous increases in food prices is that when shortages arise, there is a tendency for countries to react with "beggar-thy-neighbor" trade policies that insulate domestic consumers and producers from international price movements, and in doing so, shift the adjustment costs onto others. This has included ad hoc reductions in import tariffs and increases in export barriers, neither of which is effectively disciplined under current WTO rules. Many governments have also responded to the food crisis by focusing on measures to increase their degree of self-sufficiency in food production. In the future, as climate change makes food production increasingly high-cost in some countries, trying to maintain levels of self-sufficiency will likewise become increasingly costly. This underscores the importance of keeping the trade system open in order to give all countries confidence that they can rely on it to supply their food requirements.

Second, barriers to trade in goods and services that help reduce emissions would ideally be eliminated.

These are currently being addressed in the Doha Round negotiations, but progress has been limited. Of particular interest to LAC is the reduction of barriers to trade in ethanol. This is of greatest interest to Brazil, which is the lowest-cost producer in the world, but may be important for other countries in the Region where ethanol can be efficiently produced from sugarcane. From the dual perspectives of efficiency and effectiveness in reducing emissions, it is in the world's interest to ensure that ethanol is produced where this can be done most efficiently, rather than in countries where it requires large subsidies and high trade barriers. Current trade policies and subsidies to biofuels in high-income countries have generated huge distortions in agricultural markets, with adverse impacts on poor food consumers worldwide, and at best minimal reductions in carbon emissions.

Finally, the WTO's Committee on Technical Barriers to Trade is already involved in reviewing the increasing number of standards and labeling requirements targeted at energy efficiency or emissions control. It could also play an important role in ensuring that other trade policies—including tariffs levied on the basis of the producing country's emission reduction commitments or environmental regulations—are not discriminatory and do not unnecessarily restrict trade.

A LAC-friendly global climate change architecture is also needed

For LAC, as for other developing countries, the architecture of the post-2012 climate regime will be critical. As currently designed, the Clean Development Mechanism (CDM) cannot deliver LAC's potential to reduce its GHG emissions in a cost-effective way.[103] In the design of the post-2012 regime, there are two prominent issues for LAC. First, from the perspective of high-volume cost-effective mitigation and critical biodiversity protection, the new chapter of the regime must incorporate REDD. Second, from the perspective of long-term low-carbon (sustainable) economic growth, the Region needs a mechanism for carbon finance that goes beyond the project-based approach of the CDM in order to create incentives to significantly shift the carbon intensity of investments that will be made in the energy and transportation sectors and to take advantage of the many opportunities for increasing energy efficiency.

Incorporating REDD in the international climate architecture

The single most important issue for LAC in the negotiations over the post-2012 regime is the incorporation of REDD in the international climate change architecture. The first commitment period of the Kyoto Protocol only recognized afforestation and reforestation projects in the CDM and did not include reduced emissions achieved by means of avoided deforestation or other types of forest management in developing countries. More recent international negotiations have moved towards recognizing decreases in deforestation and forest degradation from a pre-established baseline as a source of credits and/or compensation in a post-2012 regime. One important challenge in designing such schemes is how to give credit to countries which have effectively preserved their forests and so have a very low baseline rate of deforestation.

Several types of proposals for incorporating REDD have emerged during recent years. Perhaps the main distinction between the various proposals is whether developed countries would be allowed to gain credits for their possible contributions to REDD efforts in the developing world. A large number of developing countries, including several from LAC, favor a market approach in which REDD activities would give rise to tradable credits. Other countries favor a nontradable, "fund" approach. Brazil, in particular, has established a specific "nonmarket" fund dedicated to REDD. The Amazon Fund will receive contributions from industrialized countries but those will not count towards the mitigation commitments of those countries. The Fund will award financial incentives for reductions in deforestation rates below established baselines. Other proposals have combined aspects of both market-oriented and fund-based alternatives, while also establishing financial incentives per avoided ton of CO_2.[104]

Improving the mechanisms to support low-carbon development

A number of features in the global architecture would improve its ability to provide incentives for invest-ment in low-carbon technology. First, to maintain the Region's relatively clean profile in energy generation, it is especially important that the carbon-trading architecture recognize the value of hydropower. Currently the European Union, the main buyer in the market, requires that certified emission reductions derived from hydropower projects over 20 MW must comply with the guidelines of the World Commission on Dams. In practice this requirement has added complexity to project registration and prevented the registration of all but small projects. Better incorporation of hydropower into the global mechanism could reinforce the country-level actions that also need to take place as described below.

A number of additional concerns with the current functioning of the CDM need to be addressed in order to unlock LAC's full potential to contribute to reducing emissions. One problem is that the current CDM focuses on project-level emission reductions, relative to baseline scenarios. This single-project approach makes it unlikely to "catalyze the profound and lasting changes that are necessary in the overall GHG intensities of developing countries' economies" (Figueres, Haites, and Hoyt 2005). Many of the potentially good options for reductions—especially in energy efficiency and agriculture—involve measures or investments that individually have a small effect on emissions, and consequently cannot qualify as projects or are too small to justify the transactions costs associated with the CDM, but in the aggregate are significant. A more effective approach would entail transforming the baselines themselves so as to make development pathways more carbon-friendly (Heller and Shukla 2003). In this context, rather than focus on actions at the project level, mitigation efforts in developing countries would have to shift toward promoting reforms across entire sectors—for example, energy, transport, agriculture, and forestry.

One way of implementing this is to broaden the CDM to include reductions obtained by developing countries while pursuing climate-friendly development policies. One first important step in this direction was the decision to include programs of activities in the CDM, taken in December 2005 in Montreal. This so-called "programmatic approach" could be especially relevant in the areas of energy efficiency and fossil fuel switching, where the deployment of low-carbon technologies usually occurs through multiple coordinated actions executed over time, often by a large number of households or firms, as the result of a government measure or a voluntary program. In this new approach those programs of activities—and not just the individual projects—can be made eligible for the sale of emission reduction credits, which greatly reduces transaction costs and thus facilitates the participation in the mechanism of less developed small and medium countries.

Other proposed extensions of the CDM—not yet accepted—include the so-called policy-based and sectoral approaches. The former aims to create incentives to transform overall development policies and make them more climate-friendly. Emission reduction credits would be awarded to developing countries that successfully meet nonbinding commitments to reduce GHG emissions, by means of policies and measures aimed primarily at sustainable development objectives. The first step in this direction was the decision in 2005 to include programs of activities in the CDM, but further developments are needed to enhance the impact of this mechanism. In the "sectoral" approach (Samaniego and Figueres 2002), emission reduction credits would be awarded to developing countries that overachieve on mitigation targets adopted voluntarily for specific sectors. The targets could take the form of fixed emission reductions, changes in emission intensities, or adoption of policies that result in emission reductions.

Priority domestic mitigation policies in LAC

To understand better the relative importance of mitigation policies across the various countries in the region, it is useful to group them in three different categories, depending on their total emissions: (a) large emitters, those countries that exceed 1 percent of global emissions; (b) low emitters, including those that emit less than one-thousandth of global emissions; and (c) a group in between.

As mentioned before, the largest regional emitters of GHGs are Brazil and Mexico (about 2.3 and 0.7 billion tons CO_2e per year, respectively, considering all GHGs).[105] These are the only countries in the region with CO_2e emissions exceeding 1 percent of global emissions, and they account for over 60 percent of the regional tally. Both are members of a group of large developing country emitters that are at the center of discussions regarding emission reductions. In the medium term, these two countries are likely to continue to dominate the CO_2 regional picture. Thus, most mitigation efforts in the region are likely to continue to put significant focus on these two economies. In the third group of "intermediate" emitters—composed of 11 countries: Argentina, Bolivia, Colombia, Chile, Ecuador, Guatemala, Nicaragua, Panama, Paraguay, Peru, and República Bolivariana de Venezuela—mitigation actions may also have some global effect. It is, however, a diverse group and mitigation priorities vary considerably across countries (see section 4 and annex 1).

Most other countries in the region, however, are low-carbon economies, defined as those with a carbon footprint of less than 40 million tons of CO_2e per year. Most of these also have low carbon intensities. This category includes Costa Rica, El Salvador, Honduras, Uruguay, and all Caribbean nations. Together this cohort has a total CO_2 contribution of less than a quarter billion tons of CO_2e (about 0.55 percent of global emissions). Furthermore, either because of their limited population or as a consequence of the composition of their emissions—typically dominated by the power and transport sectors and, in some cases, by modest rates of land use change—it is very unlikely that the GHG emissions of these nations will show significant changes in the future. And even if they do, the net global impact will be negligible. It is worth noting, however, that even in this group of smaller emitters, no-regrets mitigation options could represent non-negligible opportunities for tackling important development challenges while benefiting from the financial and technological support of the international community.

In setting priorities for mitigation efforts in LAC, it is reasonable to expect that the first priority will be given to the many measures that have low net costs (accounting for cobenefits), and offer large reductions, while looking for opportunities to benefit from financial flows in carbon markets. Of course, priorities will vary depending on country circumstances, but the sectors that appear to fit these criteria best across the region are (a) land use and land-use change (especially forestry), (b) energy generation, (c) transportation, and (d) energy efficiency.[106] All countries would also benefit from looking closely at their domestic policies and regulatory regimes to ensure that they provide a framework conducive to taking advantage of opportunities in the carbon market. This suggests the high priority of the policy objectives discussed in the succeeding sections.

Reduce emissions from land-use change

While it is critically important to LAC that the future climate architecture incorporate REDD activities, this is also an agenda that countries have an interest in pursuing outside the global architecture, either unilaterally or bilaterally.

Effective domestic forest policies are the cornerstone of efforts to reduce emissions from this source as well as to increase the resilience of these ecosystems to prepare them for the changing climate. Many countries in the LAC region have designed good laws and regulations in the forestry sector, but effectively implementing them and ensuring that they achieve forest conservation objectives has proved challenging. Several of the main constraints to halting deforestation are (a) the fact that politically difficult policy actions are required; (b) the need for adjustment to development strategies that go well beyond forests but impact forests (including agriculture, transportation, mining, and energy); and (c) rising population pressure.

Two prominent approaches to management of forests are protected areas and regulated concessions on privately owned land. Privately owned forests include areas managed by local communities, local governments, or individual owners. Management of a relatively small but growing share of forests in LAC is being decentralized to local governments and indigenous communities, especially since the recognition of indigenous land rights has found particularly strong resonance in this region. The share of privately owned forests in LAC by far exceeds private forest ownership in other regions, with 56 percent in Central America,

17 in South America excluding Brazil, and 15 in the Caribbean compared to the global average of 13 percent.[107] Community-based forest management in Mexico has reached a scale unmatched anywhere else in the world; an estimated three-fourths of Mexican forests are communally owned either by *ejidos* or indigenous communities.

Land tenure matters in the way forests are managed. Recent empirical comparisons of different types of forest ownership indicate that in communally owned forests, both carbon sequestration and livelihoods benefits can best be achieved if certain measures are taken. These include increasing the area of the forests under community control, giving greater autonomy to local communities in managing their forests, and compensating them to reduce forest use.[108] In other types of privately owned forests, successful innovative approaches include a shift from regulation to economic instruments such as transferable forest obligations in the Amazon in Brazil and payment for environmental services programs. Nationally managed protected areas tend to be more effective if they have sufficient staff; guards are important for transforming "paper parks" into working parks and working with local residents.[109] But too often such protected areas are underfunded, with the result that deforestation continues unabated. On the flip side, stringent enforcement may have adverse social consequences on the forest communities if regulations prohibit the use of forest products. The economic and social costs of creating parks must be weighed against the economic opportunities presented by other types of management to improve both the social outcomes and the political feasibility of forest protection measures.

Policies and large investments outside the forest sector—energy and agricultural policy, road building, and other large infrastructure projects—have a very large impact on forest resources. By opening up new forest frontiers for agricultural and logging activities, roads are the single most important driver of deforestation. Agro-ecological zoning is one of the ways to mitigate the deforestation pressure created by road construction. The participatory agro-ecological zoning process involves identification of areas of high biodiversity value and prioritization of infrastructure and other development early on in the planning process, while taking into account the economic growth and conservation objectives. Recent modeling efforts show that better road planning, agro-ecological zoning, and effective enforcement of conservation objectives in protected areas and private lands can reduce future emissions from deforestation in Brazil by half.[110]

Only a concerted, multisectoral approach can make forest conversion less attractive relative to other land-use options and reduce the pressures stemming from these sectors. But tailor-made policy solutions are needed to address particular drivers of deforestation while recognizing the specificities of each country's social and economic setting and its state of forest resources. In this regard, LAC offers a very broad range of situations: from high deforestation (for example, in Nicaragua) to net reforestation (for example, in Costa Rica) to historically low deforestation (for example, in Guyana). Oftentimes agriculture is a key deforestation driver, sometimes as a result of policy incentives to extensive cattle farming or crop cultivation. Unclear land tenure is an outstanding feature of several of the region's countries that needs to be addressed. Of particular relevance to REDD, technical and human monitoring capacity, forest management know-how, and capability vary significantly among countries within the region. Hence, a mix of customized policies is needed to address the forest-climate nexus in each of the Region's countries. Initiatives such as the Forest Carbon Partnership Facility (FCPF) of the World Bank recognize the heterogeneity by country and seek to build capacity for custom-made solutions addressing REDD (box 7).

Countries in the LAC region are the world's leaders in implementing incentive-based payment schemes for forest conservation. In 1996, Costa Rica passed the Forest Law 7575, which has recognized that forest ecosystems generate valuable ecosystem services and provided the legal basis for the owners of forest lands to sell these services. A large number of contracts were intermediated by the National Fund for Forest Financing (FONAFIFO) as a result. Most of these payments to landowners have been for hydrological

services and watershed protection—financed by such enterprises as hydropower generators and by municipalities—but availability of new financing through the CDM for afforestation and reforestation activities and payments for REDD are a promising source of revenues for Costa Rica in the future (Pagiola 2008). To a large extent, Costa Rica is now hailed as the global pioneer of payments for environmental services produced by forests. Mexico's experience with the ProArbol Program (box 8) illustrates that these programs have great potential to attract interest from land users. But to be effective they must be carefully designed with clear criteria to target payments in ways that meet the program's objectives. Conservation banking schemes (box 9) provide additional examples of the emerging innovations in this area.

Designing effective policies, however, requires good information on how land-use change affects emissions. In general, countries that are interested in moving forward with a REDD strategy may wish to consider the following steps: (a) fine-tuning the estimation of emissions from land-use change at the subnational level using high-resolution imagery (for example, Landsat with a 30-meter resolution); (b) conducting a national forest inventory to estimate carbon stocks; (c) adopting a spatially explicit modeling approach to predict future deforestation; and (d) establishing a national monitoring, reporting and verification system capable of tracking changes in deforestation and forest degradation and the resulting GHG emissions. Several LAC countries are already using or planning to use high-resolution remote sensing techniques to establish their baseline deforestation trends and monitor deforestation over time. Several forest inventories are also being planned in the countries that do not have one—few currently do, because of the cost involved.

Transform urban transport

Many "low-hanging fruits" for mitigation are available in the Region's transportation sector but few have been harvested. What are the crucial policy measures in the sector to tackle the regulatory and institutional barriers and market failures that may have prevented

BOX 7

Supporting Customized Solutions through the Forest Carbon Partnership Facility (FCPF)

The FCPF intends to build the capacity of developing countries, including at least 10 from LAC (Argentina, Bolivia, Colombia, Costa Rica, Guyana, Mexico, Nicaragua, Panama, Paraguay, and Peru), to benefit from future systems of positive incentives for REDD. As part of the capacity building, countries receive assistance to adopt or refine their national strategy for reducing emissions from deforestation and forest degradation.

The Readiness Plan Idea Notes prepared by the LAC countries participating in the FCPF so far suggest that most of their programs and activities designed to reduce emissions from deforestation and degradation will fall in the following categories: (a) general economic policies and regulations; (b) forest policies and regulations; (c) economic mechanisms for forest conservation; (d) rural development programs; and (e) social programs.

Examples of general economic policies and regulations for REDD include Guyana's willingness to promote less destructive practices in mining and road development and Mexico's efforts to mainstream forest conservation in agriculture and transportation.

Forest policies and regulations are likely to form the bulk of LAC's REDD programs and activities. Argentina, Mexico, and Nicaragua are establishing alternative forest management practices fostering the creation of economic opportunities for forest-dependent communities. Bolivia and Mexico are promoting community forestry. Colombia and Guyana favor reduced-impact logging. Costa Rica, Guyana, Mexico, Nicaragua, and Panama provide incentives for reforestation and plantations to relieve pressure on natural forests. Costa Rica and Mexico see the need to reinforce the protection and management of their system of protected areas. Several countries emphasize the need for better forest law enforcement. Paraguay wishes to decentralize forest management to empower local governments in the conservation and sustainable use of forest resources. Guyana relies on log tagging and tracking to reduce illegal logging.

Several types of economic mechanisms for forest conservation are in use or in preparation in LAC countries. Costa Rica and Mexico will continue to rely on payments for environmental services for protection, reforestation, and forest regeneration, and Colombia may start doing so. Guyana has been using forest concessions. Panama may scale up its experience with debt-for-nature swaps. Bolivia is thinking about experimenting with tradable deforestation permits.

With respect to rural development programs, Bolivia recognizes the need for silvopastoral systems as a more efficient and less destructive alternative for cattle ranching, and for the development of income-generation activities in the highlands so as to reduce migration to the lowlands of the Amazon region. Guyana proposes to foster ecotourism, handicrafts using nontimber forest products, aquaculture, and rural electrification. Panama will improve its land administration and continue to promote investment projects at the subnational level to improve rural livelihoods, while Peru is launching a number of REDD pilot projects to identify the activities that are necessary to reduce poverty.

Finally, several LAC countries are proposing a range of social programs expected to generate direct or indirect benefits in terms of REDD. Argentina proposes to confer ownership rights over forest land to indigenous and rural communities and halt the internal displacement of indigenous peoples. Bolivia wants to promote the sustainable use of nontimber forest resources, wildlife, and environment services by peasant communities and indigenous populations, according to their knowledge, uses, and customs. Guyana will engage with Amerindian communities to use their titled lands in sustainable ways. Panama will rely on the ongoing Sustainable Rural Development program of the indigenous Ngöbe Buglé Region in an effort to reduce poverty and poverty-related deforestation.

BOX 8

Paying to Protect Forests through ProÁrbol in Mexico

In 2003, Mexico instituted a program of payments for hydrological environmental services. This evolved into a broader program of payments for environmental services of forests, which in turn is part of a program of support to forests, ProÁrbol. About 1.4 million ha were under conservation contracts in early 2008; the 2008 contracts would bring this total to over 2 million ha. The program pays landowners to conserve existing forests, mainly for the services they provide in managing water resources. Payments are made ex post, after the conservation has been verified. Conservation contracts are for five years, and are conditionally renewable. Participants receive payments of about US$40/ha/yr for cloud forest and US$30/ha/yr for other forests. Although the program has grown rapidly, it was initially poorly targeted. Recent years have seen significant efforts at improving targeting by introducing clear prioritization criteria. Efforts are also underway to diversify the program away from its current one-size-fits-all approach so that it is better suited to local conditions in different parts of the country.

BOX 9

Conservation Banking to Reduce Deforestation and Protect Biodiversity

Another innovation in the region to reduce deforestation is President of Guyana Jagdeo's offer to cede the management of his country's entire rain forest (over 18 million hectares, covering more than 80 percent of Guyana's land mass) to the British government in return for economic assistance. While the offer is still on the table, the government and the 371,000-hectare Iwokrama Forest Reserve has reportedly negotiated a more limited deal with Canopy Capital, an investment group. Similar deals in other developing countries include a US$9 million investment by Merrill Lynch in Sumatra in the expectation of eventual profits from sale of carbon credits, and a "wildlife conservation banking scheme" in Malaysia established by New Forests (a Sydney-based investment firm), which expects to receive a return of 15–25 percent by selling "biodiversity credits." This underscores the potential for forests to generate financial resources even outside of the formal carbon market.

the implementation of the most promising measures with the highest mitigation potential, low costs and large cobenefits?

In contrast to most of the earlier approaches that have tended to focus on one technical or economic solution in the sector at a time, mitigation policies are more effective if they broaden the focus and simultaneously address different aspects of the transport problem: growth in private vehicle use, deteriorating public transport systems, poor nonmotorized facilities, sprawling cities, and lack of intermodal integration. This calls for comprehensive strategies that integrate transport sector and urban planning. One way to achieve this integration is through the provision of alternatives to travel in private cars, such as Bus Rapid Transit (BRT) and rail based transit systems. The region's pioneering experiences with BRTs—dedicated bus lanes, prepayment of bus fares, and efficient intermodal connections—are the entry point to a process of a broader urban transformation toward more livable cities with less congestion and better land-use planning.

The benefits from BRT and mass transit systems are magnified when combined with a broader set of land-use policies to foster densification along main transport corridors and promote intermodal integration with nonmotorized transport and other modes, including private vehicles. This set of complementary measures can reduce travel time, reduce local and global emissions, and provide other social benefits. In the case of Mexico, a combination of measures to reduce the distance of urban commuting by encouraging dense urban development, and the implementation of efficiency standards for vehicles is expected to reduce emissions over 2009–30 by, respectively 117 and 185 $MtCO_2e$, and have additional social and environmental benefits.[111] A large share of the cobenefits from more efficient public transportation systems can accrue to the poor, as is evident from the assessment of benefits distribution from time savings from the TransMilenio BRT system in Bogota (figure 13).

Apart from the provision of alternatives to the use of private vehicles, incentives for their reduced use and improved efficiency are another key element of the mitigation agenda. Any successful mitigation policy in the transport sector needs to address growth in private vehicle use and related emissions, especially in the Region's urban areas. This can be accomplished by improving fuel efficiency of vehicles and by introducing low-carbon fuels. Even more important are policies that make private vehicle use less attractive while also creating incentives for public and mass transit systems. Recent studies in Brazil have estimated that implementing improved automobile fuel efficiency standards could reduce emissions by about 25 $MtCO_2$ per year, while at the same time generating significant financial savings and reducing local pollution. In Peru, the renovation of the vehicle fleet could also lead to large emission reductions, of about 7 $MtCO_2$ per year at negative costs (considering the fuel savings). Finally, in Colombia the optimization of freight and public transport operations could make it possible to reduce emissions by 95 $MtCO_2e$ between 2007 and 2030.[112]

Reducing emissions, congestion, and local air pollution from freight transport in Latin America has emerged as another top priority on the climate policy and sector's agenda. Studies of improvements in logistics and projects to attain those improvements that are underway in the Region have identified opportunities to improve fuel efficiency and reduce GHG emissions and local air pollution at the same time.[113] Specific measures—including programs to improve operations, fleet maintenance, and driver behavior—that target major transport operators and freight companies can yield significant fuel savings, large economic benefits, and GHG emissions reductions.

Finally, making available basic data collection and assessment frameworks to decision makers and the broader set of stakeholders would improve understanding of the fundamental linkages between transport, climate change, and other economic and environmental benefits. Quantification of these cobenefits and an assessment of the feasibility of implementation is an important component of an overall evaluation of alternative—and sometimes complementary—mitigation options. The availability of cross-country information on the potential to reduce emissions in the transport sector such as this is an important contribution to facilitate the setting of priorities in sectoral mitigation policies, but estimates from the available studies are not directly comparable because of divergent and sometimes unclear assumptions. In the transport sector, these assessments need to evaluate the mitigation potential and the benefits from energy savings, reduction in local air pollution, and time savings, using consistent methodologies to ensure comparability across countries. Because of its public-good nature, the most efficient provision of this type of information in developing countries would require harmonization at the global or at least the regional level.

Transport policy decisions made in Latin America today will have a profound impact on the ability to control global greenhouse gas emissions from the sector in the future. Current policies will also in part

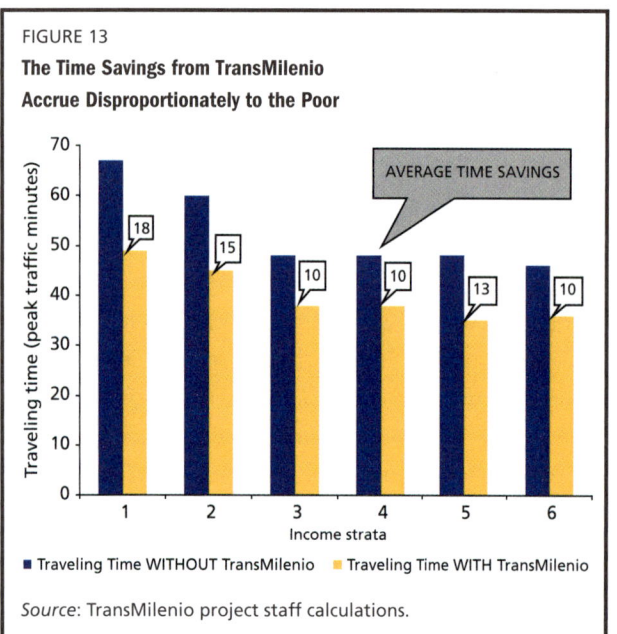

FIGURE 13

The Time Savings from TransMilenio Accrue Disproportionately to the Poor

AVERAGE TIME SAVINGS

Source: TransMilenio project staff calculations.

determine the extent to which other key development objectives, such as health outcomes, economic efficiency, and an improvement in the overall quality of life, are attained in urban areas. Implementation of policies that promote motorization—such as large-scale investments in roads and city planning that encourages urban sprawl instead of public transport systems and densification of urban areas—makes it more difficult to return to more sustainable transportation options in the future. Thus, transportation policies need to be assessed with a long-term horizon and keeping in mind that the policy options available in the future will depend on today's choices.

Continue to decarbonize growth through reliance on hydropower

Combining high-income growth—and the consequent growth in demand for electricity—with low emissions will require that LAC continue to rely on clean energy sources for a relatively large fraction of its generation capacity. The most obvious way to do this is to develop more hydropower generation, in which the Region as a whole has huge untapped potential. As noted in section 4, expansion of hydropower faces significant policy barriers, including the challenges of the licensing process. Hydropower projects can have adverse environmental and social consequences, and so are generally required to undergo some kind of licensing procedure. While the reasons for the licensing are legitimate, the process is sometimes unnecessarily long, with uncertain outcomes, and adds significantly to project costs.

Yet much has been learned and internalized about how to develop hydropower projects with minimal negative environmental and social consequences. A recent study[114] in Brazil suggested that regulatory costs could be reduced while remaining sensitive to environmental and social concerns by making a number of legislative and regulatory changes to streamline and better coordinate the process. Minimizing adverse environmental and social effects of hydropower and other clean energy projects that involve large infrastructure works requires strategic planning at the sector and subsector levels, an effective regulatory framework, environmental information, and institu-

tions that can monitor and enforce standards and regulations. Mainstreaming environmental and social considerations in project design at an early stage can significantly reduce infrastructure's environmental footprint. This can be achieved through avoiding critical natural habitats in the choice of infrastructure sites and minimizing damage to other (noncritical) natural habitats, and through such mitigation measures as careful engineering design and ecological compensation programs. Environmentally friendly options that can be considered in project design include using run of river instead of a reservoir design, or different turbine technologies for generators.

Using other instruments to complement the Environmental Impact Assessment (EIA)—including zoning plans and Strategic Environmental Assessments (SEA)—will improve infrastructure planning and the assessment of environmental impacts. The advantage of SEA is the possibility of assessing cumulative effects (for example, impacts of building several rather than one hydropower plant in the same river basin) and comparing alternatives that are not assessed in the standard EIA process. Zoning plans can also be instrumental for selecting the sites for hydropower plants and dams and helping avoid critical wildlife habitats. This approach has been successfully applied to planning roads as a network—helping avoid critical habitats and increase social benefits—in the Tocantins state in Brazil. Using these complementary tools can enhance the EIA process, improve its efficacy, and reduce regulatory costs and delays, thereby helping overcome the main obstacles to realizing the potential of the region to meet a large share of the growing energy demand from low-carbon sources.

In summary, the realities of climate change and the consequent need to reduce emissions have increased the benefits of hydropower development, while experience and advances in licensing tools have reduced the risks. In light of this, it would be useful for all stakeholders to take a new look at the cost-benefit calculus of hydropower development.

Make energy generation and use more efficient
Despite some successes, and even though most countries in LAC have already adopted a range of energy

efficiency policies, the energy savings achieved so far have been modest. Stronger public policies could provide incentives for individuals and the private sector to invest in cost-effective energy efficiency measures. While energy efficiency improvements can be undertaken one technology at a time, the best practice involves the implementation of a package of measures. And, while implementation can take place on a one-off, single-site basis, such as in a single factory or building, a far greater impact can be achieved when energy efficiency measures are implemented on a widespread, systemic basis among many users, using a combination of incentives, information, and policies to achieve the necessary market transformation. But encouraging energy efficiency is not always easy. One issue is that the party undertaking the initial investment (for example, a building owner contemplating installation of better insulation that will reduce the heating costs of tenants) may not be able to capture the benefits of the energy savings without incurring high transaction costs. Another obstacle is that reducing subsidies to energy consumption has proven to be politically sensitive. This is one reason why, in aggregate analyses, these options always seem to be "negative-cost" or "no-regrets," but are rather rare in practice. Still, a serious effort to improve energy efficiency will involve an integrated package of policies on several fronts.

The most important measures in many countries would include:

- *Encourage a switch to energy-saving technologies.* This can be done through promulgation of efficiency labeling rules, performance standards, promotion of energy efficiency among industry associations, and special programs to increase awareness of and financing for use of energy-efficient technologies.

- *Improve energy efficiency on both sides of the supply-and-demand equation for energy.* On the demand side, in addition to promotion of more efficient electrical equipment and appliances, this would include (a) supporting the creation of energy service companies to assist in identifying and financing energy efficiency opportunities in commercial and industrial consumption; (b)

promoting energy efficiency in public institutions like hospitals, schools, and government buildings through information awareness programs and changes in procurement rules to recognize the long-term savings opportunities that investments in energy-efficient products can provide; (c) demand-side management programs by electrical utilities—including changes in regulatory incentives—that encourage energy conservation and the adoption of energy-efficient practices and equipment; and (d) a reduction in electricity use by the water sector, primarily for water pumping, by reducing water losses, improving management practices, and installing more energy-efficient equipment.

- On the supply side of the equation, there are many ways to increase efficiency of electricity service provision. These include improving generation efficiency and reducing distribution losses. Several countries, including the Dominican Republic, Honduras, and Ecuador, have significant losses in distribution, through old and inefficient distribution lines and substations, as well as commercial losses stemming from theft and nonpayment. These can be improved through investments in distribution system improvement, and improved management, metering, and control. One important way to increase generation efficiency in industry and in the power sector is through cogeneration. Mexico continues to reduce carbon intensity from a high level by replacing old and inefficient plants and expanding thermal generation based on high-efficiency natural gas plants (combined-cycle gas turbines, CCGT). The energy company CFE expects that the average thermal efficiency of the group of conventional thermoelectric plants will increase from 39 percent to 46 percent during 2006–17, consistent with an increase of the participation of CCGTs in that group from 43 percent to 60 percent.

- *Reduce and better target subsidies to energy consumption.* While well-targeted subsidies are often essential for ensuring energy access by low-income or disadvantaged sectors of society, poorly-targeted fuel and electricity subsidies can lead to overconsumption of energy and increased carbon emissions. In 2005, fuel subsidies were valued at an average of 2.3 percent of GDP across the LAC region.[115] For example, Mexico and República Bolivariana de Venezuela have significant subsidies on end use of petroleum products, for example, for kerosene used in stoves or diesel in transport. Clearly, reducing these subsidies is politically difficult, but climate change provides an additional motivation, and carbon finance perhaps a source of funding to partly compensate losers and ease the transition.

Make domestic policies more carbon-trade-friendly

Countries can move on several fronts to make the local environment more conducive to development of an active market in carbon credits. A 2006 survey of investors in CDM projects found that LAC had some advantages over other regions, but slower project approvals, more host country requirements, and more differences in procedures among countries in the Region. These shortcomings could be mitigated by reducing procedural requirements and speeding up national approval processes for CDM projects. It would also be helpful for more countries to include strategies for taking advantage of the CDM in their comprehensive national climate change strategies. Currently, among countries in the Region, only Mexico and Brazil have such strategies. This would include integrating carbon-trade opportunities into sectoral strategies, for example, as potential sources of funding for projects. A related measure would be fuller participation of state-owned enterprises in the carbon markets.

6. Summary and Conclusions

Latin American and Caribbean countries are already experiencing the negative consequences of climate change. Moreover, under current trends those impacts are likely to become much more severe over the next decades. The Region's rich biodiversity, in particular, is at great risk, and agricultural productivity is likely to suffer dramatically as conditions become intolerable for current product varieties.

The impact of climate change will vary greatly across Latin American countries and subregions, not only with their level of exposure to climatic shocks, but also with their ability to adapt. Caribbean nations, for instance, are likely to be hit on multiple fronts, including through more intense natural disasters and the dieback of marine ecosystems. As a result, those nations stand to suffer relatively more, with permanent economic losses reaching by some estimates several percentage points of their GDP. Other countries will likely experience negative consequences in only some regions, for example, farmers in drought-affected areas of Brazil's northeast and water-deprived valleys of Central Chile. And, in some cases, the effects could be positive, for example, the south of Brazil and some of Chile's northern regions, which could benefit respectively from higher temperatures and increased water availability.

Because many of the climatic shocks that are likely to hit the region are to a large extent inevitable – due to inertia and the long lag times in the earth's climate system—the region's governments have to consider appropriate adaptation policies and investments. Uncertainties regarding the nature and locations of climate change impacts mean that for some kinds of responses there is value in waiting. This is true especially for investments to respond to specific effects about which the science is not yet clear (for instance, the magnitude of sea-level rise). Responses to ongoing impacts are more urgent. Fortunately, good adaptation policy is largely congruent with good development policy. In other words, many adaptive measures can be described as no regrets in the sense that they should be undertaken anyway, as part of an overall development strategy. Examples include actions to improve the region's natural resource management systems and incorporate climate related threats into the design of long-term infrastructure investments. In addition, governments can also play an important role in facilitating private responses to climate change by increasing households' flexibility and options by, for example, improving weather monitoring and forecasting; enhancing social safety nets so as to allow households to cope better with climate shocks; and enhancing the functioning of land, water, and financial markets.

Beyond adaptation policies, there is a strong case for Latin America to be an active part of a broader effort to mitigate climate change by means of drastically reducing the world's GHG emissions. As argued in this paper, for such a coordinated global mitigation effort to be effective and efficient, it must entail emission reductions also in the developing world, particularly the larger middle-income countries. Effectiveness calls for Latin American participation because even a reduction in emissions by high-income countries to zero would not suffice to keep the stock of GHG below "dangerous" thresholds. Efficiency also requires Latin American involvement because much of the low-cost, large-impact mitigation potential is located in emerging economies. However, coordinated global efforts that can engage constructive contributions by middle-income countries, including from Latin America, require a framework consistent with equity considerations—that is, a framework where the site of mitigation can be delinked from the financier of the mitigation effort and where mechanisms exist to allow countries to share the costs of climate change mitigation on the basis of their differentiated levels of "responsibility" and "capability."

Given its past record of low-carbon development, its wealth of natural resources, and its intermediate levels of income—when assessed on a global scale—many Latin American countries are well placed to take a leadership role in the developing world's response to the climate change challenge. This is not only possible; it is also in Latin America's best interest. Indeed, many of the actions needed for reducing the growth in the region's emissions are of a no-regrets nature: They would be socially advantageous regardless of their impact on climate change mitigation. In addition, adopting a low-carbon development path would benefit the Region's long-term competitiveness to the extent that the world's technological frontier moves in the direction of low-carbon technologies.

Taking advantage of these opportunities, however, requires an appropriate international policy environment in which a critical mass of high-income countries take a global leadership role. This is important not only to make such a global framework equitable, thereby lending it credibility, but also to generate sufficient incentives and momentum for the private sector to invest in low-carbon technologies. In addition, for the world to benefit from Latin America's efficient mitigation contributions, the international climate framework needs to be responsive—and welcoming—to the Region's potential contributions in the areas of forest conservation, renewable energy sources and environmentally sustainable biofuels. Finally, while taking advantage of these opportunities will require specific domestic policy actions, it is critical that the international community develop climate financing mechanisms that go beyond the project-based approach of the Kyoto Protocol's Clean Development Mechanism, and provide support to climate-friendly development policies at large.

Annex 1: Mitigation Potential by Country and Type of Emissions

TABLE A1

Relative Importance of Mitigation Potential in Energy and Non-Energy-Related Emissions Based on Emissions Growth Rates and Ratio of Emissions to GDP[116]

	Energy emissions (CO_2)	Land use change (CO_2)	Non-CO_2 emissions	Total GHG emissions in 2000 (Mt/CO_2e)
Brazil	Low	High	High	2,333
Mexico	Medium	Low	Low	682
Venezuela, R. B. de	Medium	Low	Low	384
Argentina	Medium	Low	Low	353
Colombia	Low	Low	High	274
Peru	Low	High	Medium	257
Bolivia	High	High	High	144
Chile	High	Low	Low	99
Ecuador	High	Low	Low	99
Guatemala	Medium	High	Medium	84
Nicaragua	High	High	Medium	66
Panama	Medium	High	Low	58
Paraguay	Medium	High	High	54
Guyana	Medium	High	High	39
Honduras	Medium	High	Medium	31
Dominican Republic	High	Low	Low	30
Trinidad and Tobago	Medium	Low	Medium	29
Belize	High	High	High	23
Costa Rica	Medium	Low	Low	21
Jamaica	Medium	Low	Low	16
Uruguay	Low	Low	Medium	16
El Salvador	Medium	Low	Low	15
Haiti	Low	Low	High	11
Suriname	Medium	n.a.	High	4
Antigua and Barbuda	Low	n.a.	High	2
Granada	Medium	n.a.	n.a.	0.3
Dominica	Low	n.a.	n.a.	0.2

TABLE A2

Relative Importance of Mitigation Potential in Energy-Related Emissions Based on Energy and Emissions Growth Rates and Ratio of Emissions to Energy[117]

	Energy intensity (per US$ of GDP)	Power: carbon intensity	Transport: carbon intensity	Industry and buildings: carbon intensity
Brazil	Medium	Medium	Low	Medium
Mexico	Medium	Medium	Low	Medium
Venezuela, R. B. de	High	Low	Low	Medium
Argentina	Medium	Medium	Medium	Medium
Colombia	Low	Low	Low	Medium
Peru	Low	Medium	Low	Medium
Bolivia	High	Medium	Medium	High
Chile	Low	Medium	Medium	High
Ecuador	Medium	High	Medium	Medium
Guatemala	High	High	High	Medium
Panama	Low	High	High	Medium
Paraguay	Medium	n.a.	High	Low
Honduras	Medium	High	High	Medium
Costa Rica	Medium	Medium	Medium	Low
Uruguay	Low	Low	Medium	Low
El Salvador	Medium	Medium	Medium	Medium
Haiti	High	Low	Medium	Medium

TABLE A3

Relative Importance of Mitigation Potential in Non-Energy-Related Emissions Based on Emissions Growth Rates and Ratio of Emissions to GDP[118]

	Agriculture	Waste	Other non-CO_2
Brazil	High	Low	Low
Mexico	n.a.	Medium	Medium
Venezuela, R. B. de	Low	Medium	Medium
Argentina	Low	Low	Medium
Colombia	High	High	Medium
Peru	Low	High	Medium
Bolivia	High	High	Low
Chile	Low	Low	Low
Ecuador	Low	High	Medium
Uruguay	High	Low	Low

FIGURE A1

Emissions Growth Rates and Ratio of Emissions to GDP

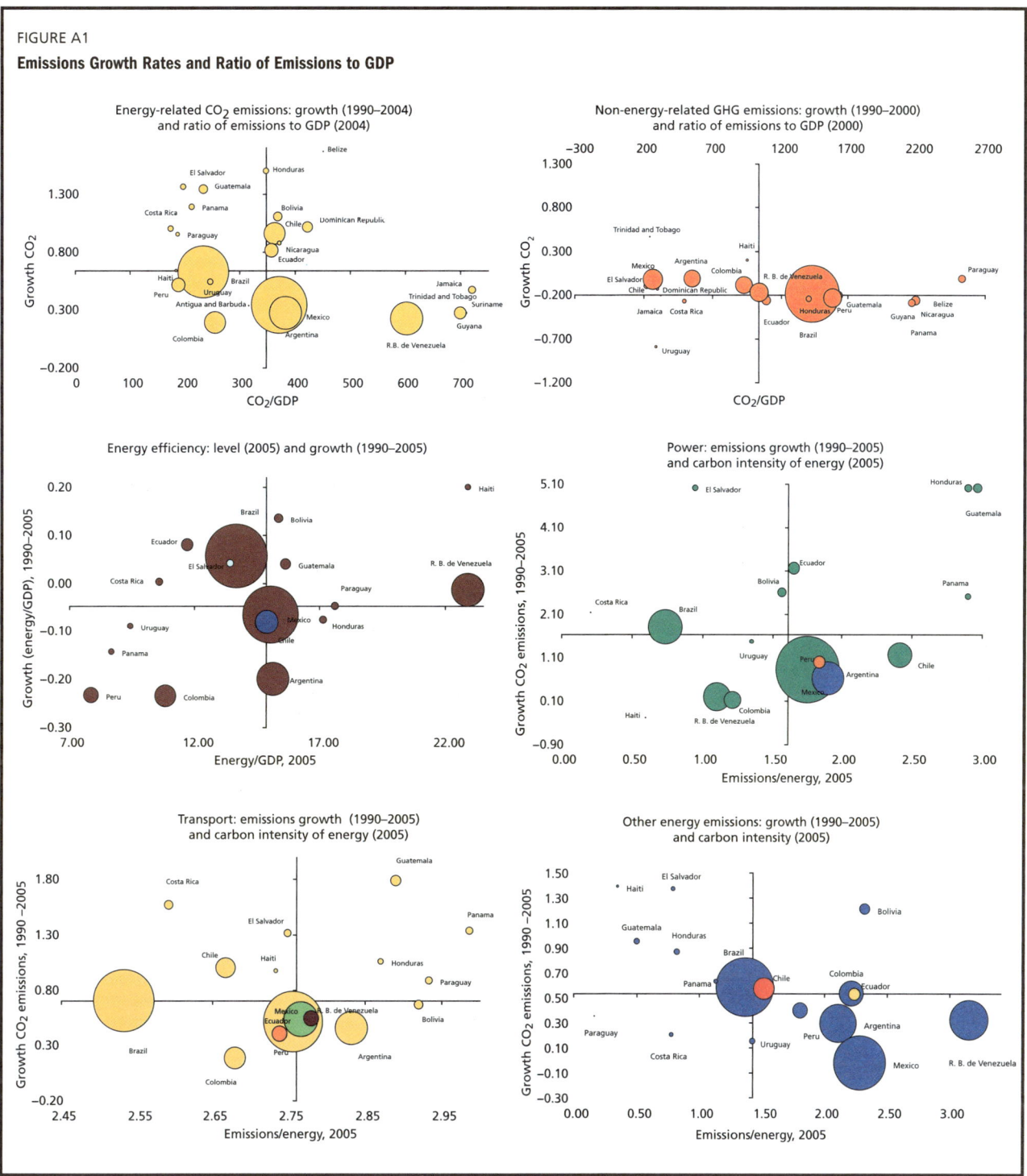

(Figure continues on next page)

FIGURE A1
(continued)

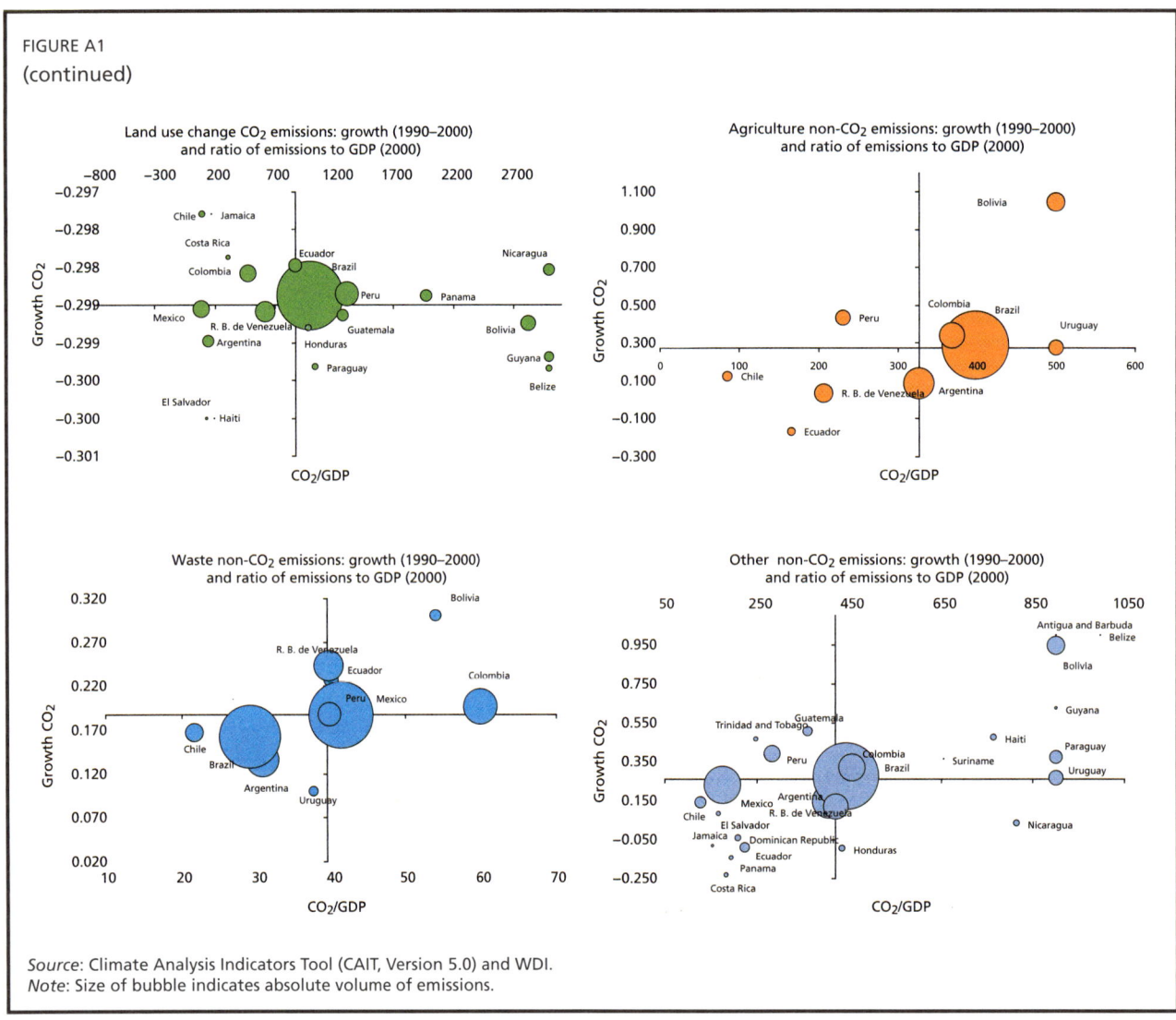

Source: Climate Analysis Indicators Tool (CAIT, Version 5.0) and WDI.
Note: Size of bubble indicates absolute volume of emissions.

Annex 2: Potential Annual Economic Impacts of Climate Change in CARICOM Countries circa 2080 (in millions 2007 US$)[119]

	Pre-subtotal	Subtotal	Total
Total GDP loss due to climate change–related disasters (hurricanes, floods):			4,939.90
Tourist expenditure		447.0	
Employment loss		58.1	
Government loss due to hurricane		81.3	
Flood damage		363.2	
Drought damage		3.8	
Wind storm damage		2,612.2	
Death (GDP/capita) due to increased hurricane-related disasters (wind storm, flood and slides)		0.1	
Floods DALY (GDP/capita)		0.8	
Sea-level rise			1,888.5
Loss of land			20.2
Loss of fish export (rising temperatures, hurricanes, and sea level)			93.8
Loss of coral reefs (rising temperatures, hurricanes, and sea level)			941.6
Hotel room replacement cost			46.1
Loss of tourists sea related tourism entertainment expenditure			88.2
Housing replacement			567.0
Electricity infrastructure loss			33.1
Telephone line infrastructure loss investment needs			3.9
Water connection infrastructure loss investment			6.7
Sanitation connection infrastructure loss investment needs			9.0
Road infrastructure loss investment needs			76.1
Rail infrastructure loss investment needs			2.7
Temperature rise			
Loss of tourists expenditure			4,027.4
General climate changes			
Agricultural loss			220.5
Water stress: cost of additional water supply			104.0
Health			
Malaria DALY (GDP/capita)			0.003
Other diseases costs			7.1
Total	**Grand total**		**11,187.30**
	% of GDP		**11.26%**

Bibliography

Agrawal, A. 2008. "Livelihoods, Carbon, and Diversity in Community Forests: Tradeoffs or Win-Wins?" Presentation at conference on "Rights, Forests, and Climate Change," October 15–17, 2008, organized by Rainforest Foundation Norway and the Rights and Resources Foundation. http://rightsandclimate.org/.

Alaimo and Lopez. 2008. *Oil Intensities and Oil Prices: Evidence for Latin America*. Washington, DC: World Bank.

Aldy J.D., Ley, E., Parry, I.W.H. 2008. "A Tax-Based Approach to Slowing Climate Change." Resources for the Future.

Alencar, A., D. Nepstad and M.D.V. Diaz. 2006. "Forest understorey fire in the Brazilian Amazon in ENSO and non-ENSO years: area burned and committed carbon emissions." *Earth Interactions* 10: 1–17.

Arnell, N.W. 2004. "Climate change and global water resources: SRES scenarios emissions and socio-economic scenarios." *Global Environmental Change* 14: 31–52.

Assad, E., and H. Silveira Pinto. 2008. "Aquecimento Global e Cenarios Futuros da Agricultura Brasileira." Mimeo, EMBRAPA and Unicamp. Available at: http://www.embrapa.br/imprensa/noticias/2008/novembro/1a-semana/site-mostra-estudo-sobre-aquecimento-global-e-seus-impactos-na-agricultura/?searchterm=silveira%20pinto.

Assuncao, Juliano J., and Flavia Fein Cheres. 2008. "Climate Change, Agricultural Productivity and Poverty." Mimeo, background paper for this report. World Bank, Washington, DC.

Avato, Patrick A. 2007. *Bioenergy: An Assessment*. Washington, DC: World Bank: Background study for the International Assessment of Agricultural Science and Technology for Development.

Avissar, R. and Werth, D. 2005. "How many realizations are needed to detect a significant change in simulations of the global climate?" Eos Trans AGU.

Bacon R., and S. Bhattacharya. 2007. "Growth and CO$_2$ Emissions: How Do Different Countries Fare?" World Bank Environment Department Papers 113, November. World Bank, Washington, DC.

Baez, J., and A. Mason. 2008. "Dealing with Climate Change: Household Risk Management and Adaptation in Latin America." Background paper for World Bank Flagship Report on Climate Change, World Bank Latin American and Caribbean Region.

Binswanger, H., and M. Rozensweig. 1993. "Wealth, Weather Risk and the Composition and Profitability of Agricultural Investments." *The Economic Journal* 103: 56–78.

Blanco, Javier T., and Diana Hernández. Forthcoming (2009). "The Costs of Climate Change in Tropical Vector-Borne Diseases—A Case Study of Malaria and Dengue in Colombia." In W. Vergara, ed., *Assessing the Consequences of Climate Destabilization in Latin America*. Sustainable Development Working Paper, LCSSD. Washington, DC: World Bank.

Boucher, O., and M.S. Reddy. 2007. "Climate trade-off between black carbon and carbon dioxide emissions." *Energy Policy*, doi:10.1016/j.enpol.2007.08.039.

Bradley, N.L., A.C. Leopold, J. Ross, and W. Huffaker. 1999. "Phenological changes reflect climate change in Wisconsin." *Proceedings of the National Academy of Sciences* 96: 9701–9704.

Bradley, R., M. Vuille, H. Diaz, and W. Vergara. 2006. "Threats to water supplies in the tropical Andes." *Science* 312: 1755.

Bresnyan, E., and P. Werbrouck. N.d. "Value Chains and Small Farmer Integration." World Bank, LCSAR, Agriculture for Development series.

Brown, J.L., S.H. Li, and N. Bhagabati. 1999. "Long-term trend toward earlier breeding in an American bird: A response to global warming?" *Proceedings of the National Academy of Sciences* 96: 5565–5569.

Buddemeier, R.W., P.L. Jokiel, K.M. Zimmerman, D.R. Lane, J.M. Carey, G.C. Bohling, and J.A. Martinich. 2008. "A modeling tool to evaluate regional coral reef responses to changes in climate and ocean chemistry." *Limnology and Oceanography Methods* 6, 395–411.

Callaway, J. M. 2004a. "Adaptation Benefits and Costs: Are they important in the global policy picture, and how can we measure them?" *Global Environmental Change* 14 (2004): 272–282.

———. 2004b. "The Benefits and Costs of Adapting to Climate Variability and Change." In *OECD, The Benefits and Costs of Climate Change Policies: Analytical and Framework Issues*. Paris: OECD.

Caso, M., I. Pisanty, and E. Ezcurra. 2004. *Diagnóstico ambiental del Golfo de México*. Vol. I y II. INE/Semarnat.

CST (Center for Sustainable Transport). 2008. "MEDEC Transport Sector." Paper commissioned for the World Bank as part of the Mexico Low-Carbon Study (MEDEC).

Charveriat, C. 2000. "Natural Disasters in Latin America and the Caribbean: An Overview of Risk." *IADB Working Paper*, No. 434, Inter-American Development Bank, Washington, DC.

Chomitz, K., et al. 2007. "At Loggerheads? Agricultural Expansion, Poverty Reduction, and Environment in the Tropical Forests." World Bank, Washington, DC.

Christensen, J.H., B. Hewitson, A. Busuioc, A. Chen, X. Gao, I. Held, R. Jones, R. K. Kolli, W.-T. Kwon, R. Laprise, V. Magaña Rueda, L. Mearns, C. G. Menéndez, J. Räisänen, A. Rinke, A. Sarr, and P. Whetton. 2007. "Regional Climate Projections." In *Climate Change 2007: The Physical Science Basis*. Contribution of Working Group I to the Fourth Assessment Report of the Intergovernmental Panel on Climate Change [Salomon, S., D. Qin, M. Manning, Z. Chen, M. Marquis, K.B. Averyt, M. Tignor, and H.L. Miller, eds.]. Cambridge, United Kingdom and New York, NY, USA: Cambridge University Press.

Cox, P.M., Betts, R.A., Collins, M., Harris P.P., Huntingford, C., Jones, C.D. 2004. "Amazonian Forest Dieback Under Climate Carbon Cycle Projections for the 21st Century." *Theor Appl Climatol* 78: 137–156.

Cox, P.M., Harris, P., Huntingford, C., Betts, R.A., Collins, M., Jones, C.D., Jupp, T.E., Marengo J., and Nobre, C. 2008. "Increasing risk of Amazonian drought due to decreasing aerosol pollution." *Nature* 453: 212–216.

Cline, W.R. 2007. "Global Warming and Agriculture: Impact Estimates by Country." Washington DC: Center for Global Development, Peterson Institute for International Economics.

Confalonieri, U., B. Menne, R. Akhtar, K.L. Ebi, M. Hauengue, R.S. Kovats, B. Revich and A. Woodward. 2007. Human Health. Climate Change 2007: Impacts, Adaptation and Vulnerability. Contribution of Working Group II to the Fourth Assessment Report of the Intergovernmental Panel on Climate Change. M.L. Parry, O.F. Canziani, J.P. Palutikof, P.J. van der Linden, and C.E. Hanson, eds. Cambridge University Press, Cambridge, UK, 391–431.

Curry, J.A.. P.J. Webster, and G.J. Holland 2006. "Mixing Politics and Science in Testing the Hypothesis That Greenhouse Warming Is Causing a Global Increase in Hurricane Intensity." *The Bull. Amer. Met. Soc.* 87(8): 1025–1037.

Curry J., M. Jelinek, B. Foskey, A. Suzuki, and P. Webster. Forthcoming (2009). "Economic impacts of hurricanes in México, Central America, and the Caribbean ca. 2020–2025." In W. Vergara, ed., *Assessing the Consequences of Climate Destabilization in Latin America.* Sustainable Development Working Paper, LCSSD. Washington, DC: World Bank.

Dasgupta, S., B. Laplante, C. Meisner, D. Wheeler, and J. Yan. 2007. "The Impact of Sea Level Rise on Developing Countries: A Comparative Analysis." Working Policy Research Paper Series 4137, February 2007.

de Gorter, H., and D.R. Just. 2008. "'Water' in the U.S. Ethanol Tax Credit and Mandate: Implications for Rectangular Deadweight Costs and the Corn-Oil Price Relationship." Paper presented at the ASSA Annual Meeting, New Orleans, Jan. 4–6, 2008.

de Gorter, H., and Y. Tsur. 2008. "Towards Genuine Sustainability Criteria for Biofuel Production." Background paper for this report, July.

De Janvry, A., Finan, F., Sadoulet, E., and Vakis, R. 2006. "Can Conditional Cash Transfers Serve as Safety Nets in Keeping Children at School and from Working When Exposed to Shocks?" *Journal of Development Economics* 79(2): 349–373.

Dunn, P.O., and D.W. Winkler. 1999. "Climate change has affected the breeding date of tree swallows throughout North America." *Proceedings of the Royal Society of London* B 266: 2487–2490.

Dussan, M. 2008. *Assessment of Climate Implications of the Energy Sector in Latin America.* The Inter-American Development Bank.

Economist, The. 2007. "Adiós to poverty, hola to consumption." August 16.

Enkvist, Per-Andres, Tomas Naucler, and Jerker Rosander. 2007. "A Cost Curve for Greenhouse Gas Reduction." *The McKinsey Quarterly* 1: 35–45.

ESMAP (Energy Sector Management Assistance Program). 2007. "LCR, Energy Sector Retrospective Review and Challenges." ESMAP, World Bank, Washington, DC.

FAO (Food and Agriculture Organization). 2005. "Global Forest Resource Assessment 2005—Progress towards sustainable forest management." In *FAO Forestry Papers* 147.

———. 2007. "Adaptation to Climate Change in agriculture, forestry, and fisheries: Perspective, Framework, and Priorities." Rome: Interdepartmental Working Group on Climate Change.

Fargione, J., J. Hill, D. Tilman, S. Polasky, and P. Hawthorne. 2008. "Land Clearing and the Biofuel Carbon Debt." *Science Express* 2: 1–3.

Farrel, Alexander E., Richard J. Plevin, Brian T. Turner, Andrew D. Jones, Michael O'Hare, and Daniel M. Kammen. 2006. "Ethanol Can Contribute to Energy and Environmental Goals." *Science* 311 (5760): 506–8.

Figueres, C. 2008. "The Global Challenge: Developing an International Climate Change Architecture." Background paper for this report.

Figueres, C., E. Haites, and E. Hoyt. 2005. "Programmatic CDM project activities: Eligibility, methodological requirements, and implementation." Washington, D.C.: World Bank Carbon Finance Unit Working Paper.

Foden, W., Mace, G., Vié, J.-C., Angulo, A., Butchart, S., DeVantier, L., Dublin, H., Gutsche, A., Stuart, S., and Turak, E. 2008. "Species susceptibility to climate change impacts." In J.-C. Vié, C. Hilton-Taylor and S.N. Stuart, eds., *The 2008 Review of The IUCN Red List of Threatened Species.* Switzerland: IUCN Gland.

Francou, B., and A. Coundrain. 2005. "Glacier shrinkage in the Andes and consequence for water resources." *Hydrology Science Journal.*

Gerolomo, M., and M.L.F. Penna. 1999. "The seventh pandemy of cholera in Brazil." *Informe Epidemiologico do Sus* 8(3): 49–58.

Giambiagi and Ronci. 2004. "Fiscal Policy and Debt Sustainability: Cardoso's Brazil, 1995–2002." IMF Working Paper 04/156. IMF, Washington, DC.

Gibbs, H., M. Johnston, J. Foley, T. Holloway, C. Monfreda, N. Ramankutty, and D. Zaks. 2008. "Carbon Payback Times for Crop-Based Biofuel Expansion in the Tropics: the Effects of Changing Yield and Technology." *Environmental Research Letters* 3: 1–10.

Gisselquist, D. J. Nash, and C. Pray. 2002. "Deregulating the Transfer of Agricultural Technology: Lessons from Bangladesh, India, Turkey, and Zimbabwe." *World Bank Research Observer* 17(2): 237–266.

Glantz, M., and D. Jamieson. 2000. "Societal response to Hurricane Mitch and intra versus intergenerational equity issues: Whose norms apply?" *Risk Analysis* 20(6): 869–882.

———. 2002. "Societal Response to Hurricane Mitch and Intra- versus Intergenerational Equity Issues: Whose Norms Should Apply?" *The Journal of Risk Analysis.*

Goland, C. 1993. "Field Scattering as Agricultural Risk Management: A Case Study from Cuyo Cuyo, Department of Puno, Peru." *Mountain Research and Development* 13(4): 317–338.

Gurgel, Angelo, John M. Reilly, and Sergey Paltsev. 2008. "Potential Land Use Implications of a Global Biofuels Industry." MIT Program on the Science and Policy of Global Change. Report No. 155.

Harris, N., S. Grimland, T. Pearson, and S. Brown. 2008. "Climate mitigation opportunities from reducing deforestation across Latin America and the Caribbean." Report to World Bank, October 2008. Winrock International.

Heller, T.C., and P.R. Shukla. 2006. "Development and climate: engaging developing countries." In *Beyond Kyoto: Advancing the International Effort against Climate Change.* Pew Center on Global Climate Change.

Hill, Jason, Nelson Erik, David Tilman, Stephen Polasky, and Douglas Tiffany. 2006. "Environmental, Economic and Energetic Costs and Benefits of Biodiesel and Ethanol Biofuels." *PNAS* 103(30): 11206–10.

Houghton. 2005a. "Above Ground Forest Biomass and the Global Carbon Balance." *Global Change Biology* 11: 945–958.

———. 2005b. "Tropical deforestation as a source of greenhouse gas emission." In *Tropical Deforestation and Climate Change*, P. Moutinho and S. Schwartzman, eds., pp. 13–22.

Howitt, R., and E. Pienaar. 2006. "Agricultural Impacts." Pp. 188–207 in J. Smith and R. Mendelsohn, eds., *The Impact of Climate Change on Regional Systems: A Comprehensive Analysis of California.* Northampton, MA: Edward Elgar Publishing.

Hoyos, C.D., P.A. Agudelo, P.J. Webster, and J.A. Curry, 2006. "Deconvolution of the factors contributing to the increase in global hurricane intensity." *Science* 312: 94–97.

Huntingford, C., R.A. Fisher, L. Mercado, B.B. Booth, S. Sitch, P. P. Harris, P. M. Cox, C. D. Jones, R. A. Betts, Y. Malhi, G. R. Harris, M. Collins, and P. Moorcroft. 2007. "Towards Quantifying Uncertainty in Predictions of Amazon 'dieback'." *Philos Trans R Soc Lond B Biol Sci.* 363(1498): 1857–1864.

Hurd, B., J. Callaway, J. Smith, and P. Kirshen. 1999. "Economics Effects of Climate Change on U.S. Water Resources." In R. Mendelsohn and J. Smith, eds., *The Impact of Climate Change on the United States Economy*, pp. 133–177. Cambridge, UK: Cambridge University Press.

IPCC (Intergovernmental Panel on Climate Change). 1996. *Climate Change 1995: Impacts, Adaptations, and Mitigation of Climate Change: Scientific-Technical Analyses. Contribution of Working Group II to the Second Assessment Report of the Intergovernmental Panel on Climate Change* [R.T. Watson, M.C. Zinyowera, and R.H. Moss, eds.] Cambridge, UK, and New York, NY, USA: Cambridge University Press, pp. 1–18.

———. 2001. *Climate Change: Impacts, Adaptation, and Vulnerability—Contribution of Working Group II to the IPCC Third Assessment Report.*

———. 2007: *Synthesis Report, An Assessment of the Intergovernmental Panel on Climate Change*, Figure 2.1 (c). Share of different sectors in total anthropogenic GHG emissions in 2004 in terms of CO_2 eq.

International Energy Agency. 2007. *World Energy Outlook.*

International Road Federation (IRF). 2006. World Road Statistics 2006. Geneva: IRF.

Kasa and Naess. 2005. "Financial Crisis and State-NGO Relations: The Case of Brazilian Amazonia, 1998–2000." *Society and Natural Resources* 18: 791–804.

Kartha, Sivan. 2006. "Environmental Effects of Bioenergy." In Peter Hazell and R.K. Pachauri, eds., *Bioenergy and Agriculture: Promises and Challenges.* Washington, DC: International Food Policy Research Institute (IFPRI).

Kaya, Y. 1990. "Impact of Carbon Dioxide Emission Control on GNP Growth: Interpretation of Proposed Scenarios." Paper presented to IPCC energy and Industry Subgroup, Response Strategies Working Group.

Knight, F. 1921. *Risk, Uncertainty and Profit.* Boston MA

Kojima, Masami, Donald Mitchell, and William Ward. 2007. "Considering trade policies for liquid biofuels." ESMAP report. World Bank, Washington, DC.

Koplow, D. 2006. "Biofuels—At What Cost?" Global Subsidies Initiative, International Institute for Sustainable Development. Geneva.

Lal, R. 2004. "Soil carbon sequestration impacts on global climate change and food security." *Science* 304: 1623–1627.

Landell-Mills, N. 2002. "Developing markets for forest environmental services: an opportunity for promoting equity while securing efficiency?" In *Carbon, Biodiversity, Conservation and Income: An Analysis of a Free-Market Approach to Land-Use Change and Forestry in Developing and Developed Countries*, I. R. Swingland, E. C. Bettelheim, J. Grace, G. T. Prance, and L. S. Saunders, eds. The Royal Society, London, 1817–1825.

Lund, J., T. Zhu, S. Tanaka, M. Jenkins. 2006. "Water Resource Impacts." In J. Smith and R. Mendelsohn, eds., *The Impact of Climate Change on Regional Systems: A Comprehensive Analysis of California.* Northampton, MA: Edward Elgar Publishing, pp. 165–187.

Magrin, G., C. Gay García, D. Cruz Choque, J.C. Giménez, A.R. Moreno, G.J. Nagy, C. Nobre and A. Villamizar. 2007. Latin America. *Climate Change 2007: Impacts, Adaptation and Vulnerability. Contribution of Working Group II to the Fourth Assessment Report of the Intergovernmental Panel on Climate Change*, M.L. Parry, O.F. Canziani, J.P. Palutikof, P.J. van der Linden and C.E. Hanson, eds., Cambridge, UK: Cambridge University Press, pp. 581–615.

Marland, G., M. Obersteiner, and B. Schlamadinger, 2007. "The carbon benefits of fuels and forests." *Science* 318: 1066.

Mata, L. J., and C. Nobre. 2006. "Impacts, Vulnerability and Adaptation to Climate Change in Latin America." Background paper for UNFCCC. 2007. Climate Change: Impacts, Vulnerabilities and Adaptation in Developing Countries. UNFCCC Report. Available at: http://unfccc.int/files/adaptation/adverse_effects_and_response_measures_art_48/application/pdf/200609_background_latin_american_wkshp.pdf.

McKinley, G.A., M. Zuk, M. Höjer, M. Avalos, I. Gonzalez, R. Iniestra, I. Laguna, M.A. Martinez, P. Osnaya, and J. Martinez. 2005. "Quantification of local and global benefits from air pollution control in Mexico City." *Envi. Sci.Technol.* 39: 1954–1961 (doi:10.1021/es035183e).

Medvedev, D., and van der Mensbrugghe. 2008. "Climate Change in Latin America: Impact and Mitigation Policy Options." The World Bank, Washington, DC.

Melillo, J.M., A.D. McGuire, D.W. Kicklighter, B. Moore, C. J. Vorosmarty, A. L. Schloss. 1993. "Global Climate Change and Terrestrial Net Primary Production." *Nature* 363: 234–240.

Mendelsohn, R. 2008a. "Impacts and Adaptation to Climate Change in Latin America", background paper for this report (September 9).

———. 2008b. "Impact of Climate Change on the Rio Bravo River." July 2. World Bank, Washington, DC.

Mendelsohn, R., ed. 2007. *The Impact of Climate Change on Regional Systems: A Comprehensive Analysis of California.* Northampton, MA: Edward Elgar Publishing, pp 165–187.

Mendelsohn, R., et al. 2008. "Long-Term Adaptation: Selecting Farm Types Across Agro-Ecological Zones in Africa." The World Bank, Washington, DC.

Mendelsohn, R., P. Christiansen, and J. Arellano-Gonzalez. 2008. "Ricardian Analysis of Mexican Farms." Background paper for this report (September 9). World Bank, Washington, DC.

Mendelsohn, R., and L. Williams. 2004. "Comparing Forecasts of the Global Impacts of Climate Change." *Mitigation and Adaptation Strategies for Global Change* 9: 315–33.

Mendelsohn, R.O., Morrison, W.N., Schlesinger, M.E., Andronova, N.G., 1998. "Country-specific market impacts of climate change." *Climatic Change* 45(3–4), 553–69.

MEDEC (2008). Mexico Low-Carbon Study (MEDEC), World Bank, 2009 (forthcoming).

Michaels, P. 2008. "Confronting the Political and Scientific Realities of Global Warming." Washington DC: Cato Institute for the Hokkaido G8 Summit.

Milly, P.C.D., K.A. Dunne, and A.V. Vecchia. 2005. "Global pattern of trends in streamflow and water availability in a changing climate." *Nature* 434: 561–562.

Mitchell, D. 2008. "A Note on Rising Food Prices." Draft, Development Economics Vice-Presidency (DECPG), The World Bank, Washington, DC.

Mueller and Osgood. 2008. "Long-term Impacts of Droughts on Labor Markets in Developing Countries: Evidence from Brazil." *Journal of Development Studies.*

Nepstad, D. C., R. E. Gullison, P. C. Frumhoff, J. G. Canadell, C. B. Field, K. Hayhoe, R. Avissar, L. M. Curran, P. Friedlingstein, C. D. Jones, C. Nobre. 2007. "Tropical Forests and Climate Policy." *Science* 316(5827).

Nordhaus, W. 2007. *The Challenge of Global Warming: Economic Models and Environmental Policy.* MIT Press.

Nordhaus W.D., and Boyer J. 2000. *Warming the World, Economic Models of Global Warming.* MIT Press.

Nyberg, J. 2007. "Sugar-Based Ethanol International Market Profile." Background paper for the Competitive Commercial Agriculture in Sub–Saharan Africa (CCAA) Study. FAO and World Bank, citing figures from UNICA. Available at: http://siteresources.worldbank.org/INTAFRICA/Resources/257994-1215457178567/Ethanol_Profile.pdf.

Pagiola, Stefano. 2008. "Payments for Environmental Services in Costa Rica." *Ecological Economics* 65(4): 712–724. Special Issue on "Payments for Environmental Services in Developing and Developed Countries," edited by Sven Wunder, Stefanie Engel, and Stefano Pagiola.

Parmesan, C. 1996. "Climate and species' range." *Nature* 382: 765–766.

Pielke R.A., Joel Gratz, Christopher W. Landsea, Douglas Collins, Mark A. Saunders, Rade Musulin. 2008. "Normalized Hurricane Damage in the United States: 1900–2005." *Natural Hazards Review*, ASCE.

Proost, S., and Denise Van Regemorter. 2003. "Interaction between local air pollution and global warming and its policy implications for Belgium." *International Journal of Global Environmental Issues* 3(3): 266–286.

Pyndick, R.S. 2007. "Uncertainty in Environmental Economics." *Review of Environmental Economics and Policy* 1(1) (winter): 45–65.

Raddatz, C. 2008. "The Macroeconomic Costs of Natural Disasters: Quantification and Policy Options." World Bank, Washington, DC.

Raupach, Michael R., Gregg Marland, Philippe Ciais, Corinne Le Quéré, Josep G. Canadell, Gernot Klepper, and Christopher B. Field. 2007. "Global and regional drivers of accelerating CO$_2$ emissions." *Proceedings of the National Academy of Sciences of the United States of America*, 04 (24) (June): 10288–10293.

Reilly, J.M., and D. Schimmelpfenning. 1999. "Agricultural impact assessment, vulnerability, the scope for adaptation." *Climate Change* 43: 745–788.

Rios Roca, A. R., M. Garron B., and P. Cisneros. 2005. "Targeting Fuel Subsidies in Latin American and the Caribbean: Analysis and Proposal." Latin American Energy Organization (OLADE), June.

Rosegrant, Mark W., Tingju Zhu, Siwa Msangi, Timothy Sulser, "The Impact of Biofuel Production on World Cereal Prices." Unpublished paper quoted with permission July 2008. International Food Policy Research Institute, Washington, DC.

Rosenzweig, M.R., and H.P. Binswanger. 1993. "Wealth, Weather Risk and The Composition and Profitability of Agricultural Investments." *Economic Journal* 103: 56–78.

Ruiz-Carrascal et al. 2008. Bi-monthly report to the World Bank on Environmental Changes in Páramo Ecosystems. LCSSD, World Bank, Washington, DC.

Ruta and Hamilton. 2008. "Environment and the global financial crisis." Mimeo, World Bank.

Samaniego, J., and C. Figueres. 2002: "Evolving to a Sector-Based CDM." Chapter 4 in *Building on the Kyoto Protocol: Options for Protecting the Climate,* Kevin Baumert, ed. World Resources Institute.

Sawyer, D. 2008. "Climate Change, Biofuels and Eco-Social Impacts in the Brazilian Amazon and Cerrado." *Philosophical Transactions of the Royal Society* 363: 1747–52.

Schlamadinger, Bernhard, Tracy Johns, Lorenzo Ciccarese, Matthias Braun, Atsushi Sato, Ahmet Senyaz, Peter Stephens, Masamichi Takahashi, and Xiaoquan Zhang. 2007. "Options for including land use in a climate agreement post-2012: improving the Kyoto Protocol approach." *Environmental Science and Policy* 10: 295–305.

Schneider, S. H., and J. Lane. 2006. "An Overview of Dangerous Climate Change." In Schellnhuber, H., ed., *Avoiding Dangerous Climate Change.* Cambridge and New York: Cambridge University Press, Chapter 2, pp. 7–23.

Schneider, S.H., S. Semenov, A. Patwardhan, I. Burton, C.H.D. Magadza, M. Oppenheimer, A.B. Pittock, A. Rahman, J.B. Smith, A. Suarez, and F. Yamin. 2007. "Assessing key vulnerabilities and the risk from climate change." *Climate Change 2007: Impacts, Adaptation and Vulnerability. Contribution of Working Group II to the Fourth Assessment Report of the Intergovernmental Panel on Climate Change,* M.L. Parry, O.F. Canziani, J.P. Palutikof, P.J. van der Linden and C.E. Hanson, Eds. Cambridge, UK: Cambridge University Press, pp. 779–810.

Searchinger, T., Heimlich, R., Houghton, R. A., Dong, F., Elobeid, A., Fabiosa, J., Tokgoz, S., Hayes, D., and Yu, T-H. 2008. "Use of U.S. Croplands for Biofuels Increases Greenhouse Gases through Emissions from Land Use Change." *Science Express* 319: 1238–1240.

Seo and Mendelsohn. 2008. "An analysis of crop choice: Adapting to climate change in South American farms." *Ecological Economics* 67: 109–116.

Smith, Pete, Daniel Martino, Zucong Cai, Daniel Gwary, Henry Janzen, Pushpam Kumar, Bruce McCarl, Stephen Ogleh, Frank O'Mara, Charles Rice, Bob Scholes, Oleg Sirotenko, Mark Howden, Tim McAllister, Genxing Pan, Vladimir Romanenkov, Uwe Schneider, and Sirintornthep Towprayoon. 2007. "Policy and technological constraints to implementation of greenhouse gas mitigation options in agriculture." *Agriculture, Ecosystems and Environment* 118: 6–28. Online at sciencedirect.com.

Smith J., and R. Mendelsohn, eds. 2006. *The Impact of Climate Change on Regional Systems: A Comprehensive Analysis of California.* Northampton, MA: Edward Elgar Publishing, pp. 188–207.

Soares-Filho, B. et al. 2006. "Modelling conservation in the Amazon basin." *Nature* 440 (23 March): 520–523.

Sohngen and Sedjo, 2006 reference, GCOMAP (Sathaye et al., 2007 reference), and IIASA-DIMA Benitez-Ponce et al., 2007 reference.

Spence et al. 2008. *The Growth Report, Strategies for Sustained Growth and Inclusive Development.* Commission on Growth and Development.

Stern, N. 2008. "The Economics of Climate Change." *American Economic Review* 98(2): 1–37.

Strzepek, K., D. Yates, and D. El Quosy. 1996. "Vulnerability assessment of water resources in Egypt to climatic change in the Nile Basin." *Climate Research* 6: 89–95.

Swiss Re. 2007. Insurance in Emerging Markets: Sound Development. Greenfield for Agricultural Insurance. Sigma No.1/2007. Zurich, Switzerland.

Szklo, A.S., Schaeffer, R., Schuller, M.E., Chandler, W. 2005. "Brazilian Energy Policies Side-Effects on CO$_2$ Emissions Reduction." *Energy Policy* 33(3): 343–64.

Thomas, Chris D., Alison Cameron, Rhys E. Green, Michel Bakkenes, Linda J. Beaumont, Yvonne C. Collingham, Barend F. N. Erasmus, Marinez Ferreira de Siqueira Alan Grainger, Lee Hannah, Lesley Hughes, Brian Huntley, Albert S. van Jaarsveld, Guy F. Midgley, Lera Miles, Miguel A. Ortega-Huerta, A. Townsend Peterson, Oliver L. Phillips, and Stephen E. Williams. 2004. "Extinction risk from climate change." *Nature* 427 (8 January): 145–48.

Toba, N. Forthcoming (2009). "Economic Impacts of Climate Change on the Caribbean Community." In W. Vergara, ed., *Assessing the Consequences of Climate Destabilization in Latin America.* Sustainable Development Working Paper, LCSSD. Washington, DC: World Bank.

Tol, R.S.J. 2002. "Estimates of the Damage Costs of Climate Change." *Environmental and Resource Economics* 21: 47–73.

———. 2005: "The Marginal Damage Costs of Carbon Dioxide Emissions: An Assessment of the Uncertainties." *Energy Policy* 33(16).

Tol, R.S.J., and G.W. Yohe. 2006. "A Review of the Stern Review." *World Economics* 7(3): 233–50.

Turner, B.T., R.J. Plevin, M. O'Hare, and A. E. Farrell. 2007. "Creating Markets for Green Biofuels." Report No. TRCS-RR-1. Berkeley: University of California. Available at: http://repositories.cdlib.org/its/tsrc/UCB-ITS-TSRC-RR-2007-1/.

UNFCCC (United Nations Framework Convention on Climate Change). 2006a. "Background paper—Impacts, vulnerability and adaptation to climate change in Latin America." UNFCCC Secretariat. Bonn, Germany: Available at: http://unfccc.int/files/adaptation/adverse_effects_and_response_measures_art_48/application/pdf/200609_background_latin_ american_wkshp.pdf.

———. 2007b. "Report on the Second Workshop on Reducing Emissions from Deforestation in Developing Countries." Available at: http://unfccc.int/resource/docs/2007/sbsta/eng/03.pdf.

Vakis, R. 2006. "Complementing Natural Disasters Management: The Role of Social Protection." Social Protection Discussion Paper, No. 0543. World Bank, Washington, DC.

Vakis, R., Kruger, D., and Mason, A. 2004. "Shocks and Coffee: Lessons from Nicaragua." Draft, Human Development Department, Latin America and the Caribbean Region, World Bank, Washington, DC.

Van Lieshout, M., R.S. Kovats, M.T.J. Livermore, and P. Martens. 2004. "Climate change and malaria: analysis of the SRES climate and socio-economic scenarios." *Global Environ. Chang.* 14: 87–99.

Vardy F. 2008. *Preventing International Crises: A Global Public Goods Perspective*. Washington, DC: World Bank.

Vasquez-Leon et al. and Conde, C., and H. Eakin. 2003. "Adaptation to climatic variability and change in Tlaxcala, Mexico." In J. Smith, R. Klein, and S. Huq, eds., *Climate Change, Adaptive Capacity and Development*. London: Imperial College Press.

Vergara, W. 2005. "Adapting to Climate Change, Lessons Learned, Work in Progress and Proposed Next Steps for The World Bank in Latin America." Latin America and Caribbean Region, World Bank, Washington, DC.

Vergara, W., A.M. Deeb, A.M. Valencia, R.S. Bradley, B. Francou, A. Zarzar, A. Grunwaldt, and S.M. Hausseling. 2007. "Economic Impacts of Rapid Glacier Retreat in the Andes." *EOS Transactions American Geophysical Union* 88: 261-268.

Vergara, Walter, Natsuko Toba, Daniel Mira-Salama, and Alehandro Deeb. Forthcoming (2009). "The consequences of climate-induced coral loss in the Caribbean by 2050-2080." In *Assessing the Potential Consequences of Climate Destabilization in Latin America*. Sustainable Development Working Paper. World Bank, Washington, DC.

Webster, P. J., G. J. Holland J. A. Curry, and H.-R. Chang. 2005. "Changes in Tropical Cyclone Number, Duration, and Intensity in a Warming Environment." *Science* 309(5742): 1844–46.

Weitzman, M.L. 2007. "A Review of the Stern Review on the Economics of Climate Change." *Journal of Economic Literature.*

West, J. M., and R. V. Salm. 2003. "Resistance and Resilience to Coral Bleaching: Implications for Coral Reed Conservation and Management." *Conservation Biology* 17(4): 956–67.

World Bank. 2008. "Environmental Licensing for Hydroelectric Projects in Brazil: A Contribution to the Debate. Brazil Country Management Unit." Report 40995-BR. World Bank, Washington, DC.

———. Forthcoming (2009). *Mexico Low-Carbon Study (MEDEC)*. LCSSD. Washington, DC: World Bank.

Worldwatch Institute. 2006. *Biofuels for Transportation. Global Potential and Implications for Sustainable Agriculture and Energy in the 21st Century.* Washington, DC: Worldwatch Institute.

Yamin, Farhana, and Eric Haites. 2006. "The São Paulo Proposal for an Agreement on Future International Climate Policy." Discussion Paper for COP-12 & COP/MOP 2.

Yamin F., J.B. Smith, and I. Burton. 2006. "Perspectives on Dangerous Anthropogenic Interference; or How to Operationalize Article 2 o f the UN Framework Convention on Climate Change." In Schellnhuber, H., ed., *Avoiding Dangerous Climate Change*. Cambridge and New York: Cambridge University Press, Chapter 2, pp. 82–91.

Zah, R., H. Boni, M. Gauch, R. Hischier, M. Lehmann, and P. Wager. 2007. "Life Cycle Assessment of Energy Products: Environmental Impact Assessment of Biofuels." Executive Summary. Mimeographed. Empa, St. Gallen, Switzerland.

Zomer, R. J., Trabucco, A., van Straaten, O., Vercot, L.V., and Muys, B. 2005. "ENCOFOR CDM-AR Online Analysis Tool: Implications of forest definition on land area eligible for CDM-AR." Published online: www://csi.cgiar.org/encofor/forest/.

Zomer, R.J., Trabucco, A., Bossio, D.C., and Verchot, L.V. 2008. "Climate Change Mitigation: A Spatial Analysis of Global Land Suitability for Clean Development Mechanism Afforestation and Reforestation." *Agricultural Ecosystems and Environment* 126(1–2): 67–80.

Endnotes

1. See, for example, Ruta and Hamilton (2008), "Environment and the global financial crisis." Mimeo, the World Bank.

2. Giambiagi and Ronci (2004), "Fiscal Policy and Debt Sustainability: Cardoso's Brazil, 1995-2002," IMF Working Paper 04/156.

3. See Kasa and Naess (2005), "Financial Crisis and State-NGO Relations: The Case of Brazilian Amazonia, 1998–2000," *Society and Natural Resources* 18: 791–804

4. "Fourth Assessment of the IPCC" (2007). The report was published in September 2007 and was produced by more than 450 authors from more than 130 countries, with more than 2,500 expert reviewers.

5. The most important anthropogenic GHG is carbon dioxide (CO_2), which in 2004 represented 77 percent of total GHG emissions. Other important GHGs are methane (CH_4) and nitrous oxide (N_2O). Global atmospheric concentrations of CO_2 increased by 35 percent between 1750 and 2005, while those of CH_4 and nitrous oxide N_2O increased by 148 percent and 18 percent respectively, during the same period.

6. Francou et al. (2005).

7. In 2004, CO_2 emissions from fossil fuel use represented 56.6 percent of total GHG emissions, while CO_2 emissions from land-use change were 17.3 percent. Agriculture was responsible for 13.5 percent of total GHG emissions, accounting for almost 90 percent of N_2O emissions (which in turn were 8 percent of total GHG emissions) and for more than 40 percent of CH_4 emissions (which were 14 percent of total GHG emissions). Other sources of CH_4 include emissions from landfill waste, wastewater, and the production and use of bio energy. IPCC (2007).

8. These concentration levels are expressed in terms of "CO_2-equivalent" units. That is, they are weighted averages of the stocks of all GHG, with weights determined by the relative warming potential of each gas with respect to CO_2. Hereafter these units will be referred to as CO_2-equivalent parts per million, or "CO_2e ppm."

9. The figure depicts observed global CO_2 emissions, from both the EIA (Energy Information Administration of the U.S. Department of Energy) (1980–2004) and global CDIAC (Carbon Dioxide Information Analysis Center of the U.S. Department of Energy) (1751–2005) data, compared with emissions scenarios and stabilization trajectories. EIA emissions data are normalized to the same mean as CDIAC data for 1990–99. The 2004 and 2005 points in the CDIAC dataset are provisional. The six IPCC scenarios are spline fits to projections (initialized with observations for 1990) of possible future emissions for four scenario families: A1, A2, B1, and B2. Three variants of the A1 (globalized, economically oriented) scenario lead to different emissions trajectories: A1FI (intensive dependence on fossil fuels), A1T (alternative technologies largely replace fossil fuels), and A1B (balanced energy supply between fossil fuels and alternatives). *The curves shown for scenarios are averages over available individual scenarios in each of the six scenario families, and differ slightly from "marker" scenarios.* The stabilization trajectories are spline fits approximating the average from two models that give similar results. They include uncertainty because the emissions pathway to a given stabilization target is not unique.

10. Magrin et al. (2007).

11. See Bradley et al. (2006). The evidence is based on analysis of ensemble analyses from global circulation models, and other analyses of field data confirm this trend.

12. National Communications to the UNFCCC (2001, 2004, 2007).

13. Caso et al. (2004). Wetlands in the Gulf of Mexico have been identified by the Mexican National Institute of Ecology (INE) as one of the most critical and threatened ecosystems by anticipated climate changes. Data published on projected forced hydroclimatic changes, as part of IPCC assessments (Milly et al., 2005) indicate that Mexico may experience significant decreases in runoffs, of the order of minus 10 to 20 percent nationally, and up to 40 percent over the Gulf Coast wetlands, as a result of global climate change. This has been documented in Mexico's third national communication to the UNFCCC.

14. These results are based on a VAR analysis for the sample of countries that have experienced at least one disaster since 1950, excluding those cases in which disasters affected less than 0.05 percent of the countries' population or GDP. See Raddtaz (2008).

15. *Notes*: Group of countries include Anguilla; Antigua and Barbuda; Argentina; Bahamas; Barbados; Belize; Bolivia; Brazil; Cayman Islands; Chile; Colombia; Costa Rica; Cuba; Dominica; Dominican Republic; Ecuador; El Salvador; French Guiana; Grenada; Guadeloupe; Guatemala; Guyana; Haiti; Honduras; Jamaica; Martinique; Mexico; Montserrat; Netherlands Antilles; Nicaragua; Panama; Paraguay; Peru; Puerto Rico; St. Kitts and Nevis; St. Lucia; St. Vincent and The Grenadines; Suriname; Trinidad and Tobago; Turks and Caicos Islands; Uruguay; República Bolivariana de Venezuela; Virgin Islands (UK); Virgin Islands (U.S.). It includes disasters that meet at least one of the following criteria: (a) 10 or more people reported dead, (b) 100 people reported affected, (c) declaration of a state of emergency, (d) call for international assistance.

16. Christensen et al. (2007).

17. There are estimates of up to a 90 percent reduction in rainfall by the end of the century (Cox 2004, 2007). However, some estimates suggest that 40 percent reductions in rainfall would suffice to initiate a dieback process.

18. According to the 2005 FAO Global Forest Resource Assessment, Latin America accounts for about 33 percent of the world's forest biomass. Moreover, estimates by Houghton (2005) suggest that the region contains 50 percent of the world's tropical forests and 65 percent of the tropical forest bio-

mass. *Global Change Biology* 11, pp. 945-958, "Above Ground Forest Biomass and the Global Carbon Balance."

19. http://www.usaid.gov/locations/latin_america_caribbean/issues/biodiversity_issue.html.

20. IPCC 2007, Thomas et al. 2004

21. The antbirds are a large family, *Thamnophilidae*, of passerine birds found across subtropical and tropical Central and South America, from Mexico to Argentina. The *Formicariidae*, formicariids, or ground antbirds are a family of smallish passerine birds of subtropical and tropical Central and South America. Manakins are found from southern Mexico to northern Argentina, Paraguay, and southern Brazil, and on Trinidad and Tobago as well. Most species live in humid tropical lowlands, with a few in dry forests, river forests, and the subtropical Andes. Source: Wikipedia.org.

22. Mendelsohn (2008a).

23. Seo and Mendelsohn (2008).

24. Mendelsohn et al. (2008b).

25. Mendelsohn and Williams (2003).

26. Tol (2002).

27. Medvedev and van der Mensbrugghe (2008).

28. The use of a discount rate of 5.5 percent is consistent with Nordhaus (2007), *Journal of Economic Literature* XLV (September 2007), pp. 686–702, "A Review of the Stern Review on the Economics of Climate Change."

29. The methodology is only applied to countries where complete economic data are readily available, specifically: Antigua and Barbuda, Barbados, Bahamas, Belize, British Virgin Islands, Cuba, Dominica, Dominican Republic, Haiti, Grenada, Honduras, Jamaica, Mexico, Nicaragua, Puerto Rico, St. Kitts and Nevis, St. Lucia, and the Grenadines.

30. Toba, N., forthcoming, 2008, "Economic Impacts of Climate Change on the Caribbean Community," in W. Vergara, ed., *Assessing the Consequences of Climate Destabilization in Latin America.*

31. If one includes Mexico in the set of affected countries, the estimated losses fall to between 0.5 and 1.2 percent of GDP. Estimates are based on the Coral Mortality and Bleaching Output model (COMBO), developed by Budenmeier and coworkers (Buddemeier et al. 2008). COMBO models the response of coral growth to changes in sea surface temperature (SST), atmospheric CO_2 concentrations, and high-temperature-related bleaching events. The model estimates the growth and mortality of corals over time based on future climate predictions and on the probability and effects of short-timed, high-temperature-related bleaching events taking place in the area. Buddemeier, R.W., Jokiel, P.L., Zimmerman, K.M., Lane, D.R., Carey, J.M., Bohling G.C. (2008). *Limnology and Oceanography Methods* 6, 395–411.

32. Javier T. Blanco and Diana Hernández, "The Costs of Climate Change in Tropical Vector-Borne Diseases—A Case Study of Malaria and Dengue in Colombia," in W. Vergara, ed., *Assessing the Consequences of Climate Destabilization in Latin America.*

33. Van Lieshout et. al (2004).

34. Gerolomo and Penna (1999).

35. The so-called greenhouse effect can be briefly described as follows: The earth's global mean climate is determined by the balance of incoming and outgoing energy in the atmosphere. Most of the energy that the earth receives from the sun is absorbed by the planet, but a fraction is reflected back into space. The amount of energy that is bounced back depends on the concentration of so-called greenhouse gases (GHGs) in the earth's atmosphere. These gases trap some of the radiation received from the sun and allow the planet's temperature to be about $30°$ C above what it would be otherwise (Stern 2007). While the greenhouse effect is a natural process without which the planet would probably be too cold to support life, the concentration of GHGs in the atmosphere has been accelerating over the past 250 years. According to IPCC (2007), there is a 95 percent probability that increases in GHG concentrations are responsible for the increases in average global temperatures and other climate trends observed over the past century.

36. Tradeoffs are mostly related to the possibility that mitigation expenditures crowd out the resources available for adaptation or possibly vice versa. Tol and Yohe (2007), for example, report that in the case of Sub-Saharan Africa the total value of expected nonmarket climate damages is highest in the most ambitious mitigation scenario, mainly because mitigation crowds out public health care. As for synergies, they are mainly derived from the fact that successful global mitigation efforts should in principle reduce the need for adaptation investments—for example, by successfully reducing the rate of global warming through reductions in GHG concentrations. In addition, some climate mitigation efforts may also increase the ability of natural and human systems to adapt to climate change impacts. Efforts to reduce deforestation for example may also foster more climate-resilient sustainable development. See, for instance, Lal (2004) and Landell-Mills (2002).

37. The optimal level of adaptation depends on the comparison of the expected damages of climate change with and without adaptive responses, as well as the costs of those responses, and the costs associated with misadapting—that is, undertaking adaptive responses in a scenario in which climate change impacts do not materialize. See Callaway (2007).

38. To see why a curve showing the marginal damages as a function of emission reductions undertaken in the present is downward sloping, consider two possible points on the curve and assume that in the future the world will implement little or no additional emission reductions (i.e., the whole curve is drawn assuming the same "business-as-usual" path for future emissions). The first point (which would be on the far left of the curve) would indicate no effort to reduce emissions from current levels. Using Stern's (2008) predictions, the earth could

eventually face a 50 percent chance of global warming in excess of 5°C, which in turn would imply a large probability of very large damages. Thus, starting from this point on the left-hand side of the curve, marginal emission reductions could have large benefits—assuming that they could allow for avoiding some of those very large damages. In contrast, starting from a point towards the right-hand side of the curve—for example, assuming that the world implements large-scale emission reductions at least on a once-and-for-all basis—it is safe to assume that the most catastrophic potential damages will at least be postponed, which implies that the marginal benefit of additional emission reductions would be smaller (at least if one assumes a positive discount rate).

39. See Vardy (2008).

40. See Knight, F. (1921). *Risk, Uncertainty and Profit.* Boston MA: Houghton Mifflin.

41. To illustrate the difficulties associated with climatic predictions, it is useful to briefly consider all the steps that are inevitably involved. One has first to deal with estimating long-run global demographic and economic trends so as to predict future *flows* and *stocks* of man-made GHG emissions—with the leap from the former to the latter involving nontrivial scientific challenges associated with the so-called carbon-cycle. Next, one has to estimate the impact that increasing stocks of GHG will have on average global temperatures and other critical climate parameters. Finally, one has to translate expected global changes in climate into regional scenarios and assess what the corresponding impacts will be on specific human and natural systems. Once again, this requires an enormous modeling effort and massive data gathering, and in the end will still leave much uncertainty.

42. See Schneider and Lane (2007) and Yamin, Smith and Burton (2007).

43. Under the UNFCCC framework, the 1997 Kyoto Protocol established a binding commitment by industrialized countries to reduce GHG emissions during 2008–12 by 5 percent with respect to their 1990 level. The Protocol was subsequently ratified by 162 countries, although some key countries, including the United States, failed to do so. The current challenge is that of reaching a follow-up agreement that, given the more recent scientific evidence, would have to extend Kyoto both in terms of the ambition of its goals and in its global coverage.

44. This measures the expected temperature increase associated with a doubling of GHG concentrations.

45. Alternatively, in a scenario where, as suggested by Stern (2008) all countries in the world would agree to converge to a common level of per capita emissions by 2050, industrialized countries would have to reduce their per capita GHG emissions to between 23 and 34 percent of their 2000 level, while developing countries would need to reduce theirs to between 64 and 96 percent of their 2000 level.

46. For the less stringent target of stabilization at 535 to 590ppm CO_2e, IPCC reports a median carbon price of 45 US$/$tCO_2$e in 2030, with model estimates ranging from 18 to 79 US$/$tCO_2$e in that year, and from 30 to 155 US$/$tCO_2$e in 2050.

47. According to IPCC, increases in energy efficiency in buildings would account for between one-fifth and one-third of global mitigation potentials. In addition, energy supply, industry, and agriculture would each account for between 15 percent and 20 percent of the total potential, while forestry could contribute 8 percent to 14 percent depending on the scenario. Emission reductions in the transport sector would account for less than 10 percent and waste for about 3 percent of the total global mitigation potential.

48. Medvedev D. and D. van der Mensbrugghe (2008). The simulations performed are, respectively, a uniform global carbon tax and a set of country-specific carbon taxes—for example, with higher taxes in countries with lower potential so as to reach the same 55 percent emission reduction in each and all countries.

49. The difference between both groups of countries is smaller but still significant when not only emissions from energy but also from land-use change are considered for the shorter 1950–2000 period. Land-use change emissions are not available from this source for previous periods. In this case, the cumulative emissions of industrialized countries would be 457 tCO_2 p/c compared to 103 tCO_2 p/c for developing countries. Data are from WRI (2008): http://cait.wri.org/cait.php (September 9, 2008).

50. In the case of *Brazil*, in October 2008 the Minister of the Environment announced that the country could achieve a 10–20 percent reduction of emissions from 2004 during the period 2012–20, presumably by reducing illegal deforestation rates. However, the government warned that these reductions are conditional on certain international prerequisites, which the Brazilian government will announce at a later date. Similarly, *Mexico*'s 2007 National Strategy on Climate Change (Estrategia Nacional de Cambio Climatico, Secretaria de Medio Ambiente y Recursos Naturales, Mexico, 2007) acknowledges the importance of urgent and concerted action on climate change mitigation and adaptation. The strategy emphasizes Mexico's willingness to engage in a more ambitious climate change framework than that established by the Kyoto Protocol and its willingness to adopt long-term targets of a nonbinding nature. The two sectors targeted for mitigation efforts are energy and land-use change and forestry. The 2007 strategy identifies a total mitigation potential of 107 Mtons in the energy sector by 2014 (representing a 21 percent reduction from BAU over the next six years) from end-use energy efficiency, increase in the use of natural gas, and increase in the cogeneration potential in the cement, steel, and sugar industries. However the bulk of Mexico's mitigation potential comes

from the land-use sector. The strategy identifies a mitigation potential that ranges from 11 to 21 billion tons CO_2 in the land-use and forestry sector by 2012, most of which will come from public reforestation and private planting, and will depend on the level of available resources. Outside of LAC, *China* is already implementing a wide range of energy and industrial policies that, while not driven by climate change concerns, are contributing to climate efforts by slowing the growth of China's GHGs. China's 11th Five-Year Plan includes a major program to improve energy efficiency nationwide, including a goal of reducing energy intensity (energy consumption per unit of GDP) by 20 percent below 2005 levels by 2010. The government projects that meeting this target would reduce China's GHG emissions 10 percent below business as usual; researchers estimate that over 1.5 billion tons of CO_2 reductions would be achieved (Pew Center for Climate Change, Climate Change Mitigation Measures in the People's Republic of China, International Brief 1, April 2007). In the case of *India*, In June 2008, Prime Minister Singh released the country's first National Action Plan on Climate Change (NAPCC), outlining existing and future policies and programs addressing climate mitigation and adaptation. The plan identifies eight core "national missions" running through 2017 and directs ministries to submit detailed implementation plans to the Prime Minister's Council on Climate Change by December 2008 (http://www.pewclimate.org/ international/country-policies/india-climate-plan-summary/ 06-2008). Emphasizing the overriding priority of maintaining high economic growth rates to raise living standards, the plan "identifies measures that promote our development objectives while also yielding co-benefits for addressing climate change effectively." The missions include: tripling renewables to 10 percent of installed capacity by 2012; 500 percent increase in nuclear power (to 20GW) by 2020; decreasing 7 percent of coal plants by 2012 and another 10,000MW by 2017, and increasing energy efficiency in order to save 10,000 MW by 2012. In South Africa, in July 2008 the government approved a progressive policy on climate change that puts the country on a low-carbon economic development path (Long Term Mitigation Scenarios: Strategic Options for South Africa, Department of Environmental Affairs and Tourism, Pretoria, South Africa, 2007). The policy calls for emissions to peak at 546 megatons of carbon by 2025 and decline in absolute terms by 2030–35. One of the measures being considered is a carbon tax, introduced by the Minister of Finance in his Budget Speech in February 2008. The Cabinet has mandated the National Treasury to study further a carbon tax as a potential option. Other measures being considered are stringent vehicle fuel efficiency standards, the development of 10,000 GWh of energy from renewable energy sources by 2012, mandatory use of carbon capture and storage (CCS) for all new coal-fired power stations, and an increase in nuclear generation. Finally, while *South Korea* has not formalized its

post-2012 intent in written form, in August 2008 Ambassador Rae-Kwon Chung, chief climate negotiator for the country, announced that South Korea would adopt a national carbon reduction target next year. A few months later he called for the establishment of an international registry for developing countries to record their domestic emission reduction policies. Registering would be voluntary, but laying out a domestic policy would translate into an international commitment that could be monitored and verified.

51. Data on tropical forest biomass are from Houghton (2005), based on 2000 FAO data. Data on share in total forest biomass are from the FAO's 2005 *Global Forest Resource Assessment*.

52. Data from the International Energy Agency.

53. Figure 9 follows the approach proposed by Kaya (1990) to decompose fossil fuel CO_2 emissions into the following factors: (a) the change in the carbon intensity of energy (emissions per unit of energy); (b) the change in the energy intensity of output (energy consumed per unit of GDP); (c) the change in GDP per capita; and (d) the change in population. Although the "Kaya decomposition" is not based on an estimated model of causal links between the relevant variables, it can be useful for uncovering the main factors driving observed changes in CO_2 emissions (see Bacon and Bhattacharya 2007). The figure reports the changes in fossil fuel emissions that can be attributed to different factors, expressed as percentage of initial 1980 levels. The figure shows that, during the past 25 years, changes in LAC's energy intensity of output contributed to increasing emissions by 15 percent, but the region's falling carbon intensity acted to reduce emissions by 17 percent. In contrast, at the global level, falling energy intensities contributed to reducing emissions by 35 percent, and reductions in carbon intensities helped reduce emissions by about 9 percent. Finally, LAC's relatively low rates of growth of per capita GDP are reflected in a smaller contribution of this factor to fossil fuel emissions, equivalent to 23 percent of the initial level, compared to 82 at the global level, 51 percent in the case of high-income countries, and as much as 309 percent in China and India.

54. As shown by Alaimo and Lopez (2008), in contrast with the evidence for the OECD, the oil and energy intensities of Latin American countries (excluding oil exporters) have not been affected by higher oil prices. To use a more technical lexicon, they are not "Granger-caused" by higher oil prices.

55. The main messages for the group of seven largest emitters are as follows: First, among countries with either high levels or high growth rates of energy related emissions, high levels of energy consumption per unit of GDP (energy efficiency) are a special concern in República Bolivariana de Venezuela, while relatively high emissions per unit of energy could be a bigger concern for Mexico, Argentina, and Chile. In Chile in particular, emissions are relatively high and growing at a fast pace in the industry and building sectors. Second, outside of energy,

land-use change is particularly important for Brazil and Peru, emissions from agriculture are either high or growing fast in Brazil and Colombia, and emissions from waste should be of special concern in Colombia and Peru.

56. World Energy Outlook (2006).

57. The study looked at the cost of reducing electricity use by 143,000 GWh in 2018 using widely available energy efficiency measures at a cost of US$16 billion compared to the costs of around US$53 billion to build the equivalent of 328 gas-powered open cycle generators (250 MW each) necessary to produce the same 143,000 GWh of power.

58. World Bank (2009).

59. Presentations made at CEPAL (Santiago de Chile) on October 16, 2008, by representatives of Fundacion Bariloche, Universidad de Chile, PSR/COPPE, Universidad de los Andes, and Universidad Catolica del Peru.

60. In addition, the opportunity to earn future carbon finance payments can increase the value of formerly marginal lands. Higher land rents improve the economic position of landowners and enhance their adaptive capacity (Lal 2004). Moreover, positive spillover effects for timber and nontimber forest products exist when sustainable forest exploitation is permitted on top of the delivery of environmental services (Landell-Mills 2002).

61. Potential land availability and location for A/R projects by country within the LAC Region were obtained by applying the ENCOFOR CDM-AR Online Analysis Tool (Zomer et al. 2008) to the crown cover threshold defined by each country under the Kyoto Protocol. This tool is available online at http://csi.cgiar.org/encofor/forest/.

62. This third group of studies models the forestry together with other sectors (agriculture and in some cases also energy) and they end up deriving supply curves. See, for instance, Boucher and Reddy (2007).

63. Expected deforestation rates, in particular, are based on multiple variables, including current deforestation trends, drivers of land-use change (roads and population growth) and land-use alternatives among others, while carbon content is determined by a series of assumptions about vegetation type and carbon pools.

64. International Road Federation (IRF). 2006. *World Road Statistics 2006*. Geneva: IRF.

65. World Bank (2009).

66. *The Economist*, 2007. "Adiós to poverty, hola to consumption," August 16th 2007.

67. http://www.time.com/time/world/article/0,8599,17338 72,00.html.

68. Estimates range from between 30 and 50 percent, according to Burtaw et al. (2003) and Proost and Regemorter (2003), to three to four times greater than total mitigation costs (Aunana, et al. 2004; McKinley et al. 2005), depending on the stringency of the mitigation level, the source sector, and

the measure and the monetary value attributed to mortality risks.

69. Aunana, et al. (2004); McKinley et al. (2005). These deaths are avoided because of a reduction in air pollution, including emissions of SO_2, N_2O, and particulate matter from vehicles and heat and power sources.

70. Mexico's energy agency, CFE, has estimated the feasible potential of wind at between 7 to 12 GW, in comparison to the current installed capacity of 51 GW, with detailed wind resource studies completed for Baja Peninsula (1500–2500MW) and the Isthmus of Tehuantepec centered in Oaxaca (2000–3000MW).

71. The wind projects in question would be those projects with high-capacity factors (about 37 percent). It is important to note, however, that the economic evaluation of generation alternatives is much more complex than the simplified analysis above based on levelized costs. One should also consider factors such as transmission costs related to the connection of the project to the national grid; local differences in operation costs and the reliability of the interconnected power system; fuel price and demand risks; externalities like the environmental impact of the projects; and fuel transportation and storage costs. From a private point of view, the economic evaluation has also to take into account the capital cost of private companies; the project, market, and country risks; costs of the firm's fuel supply; financial and fiscal incentives; transaction costs; connection and transmission costs; and power market rules and prices. See Dussan (2008).

72. Dussan (2008). The low-cost hydroelectric projects considered have investment costs below US$1,200/kW. Levelized generation costs cover fixed and variable costs, thereby including investments and operation and maintenance expenditures. The generation costs of thermoelectric alternatives vary from 41 to 65 US$/MWh for coal-fired plants; from US$49 to US$83/MWh for gas-fired plants (except for Peru, in which the cost is estimated at US$29.4/MWh and Colombia in the scenario of low oil and gas prices, for which the cost would be US$35.5/MWh); and from US$88 to US$132/MWh for diesel-fired plants.

73. Presentations made at CEPAL (Santiago de Chile) on October 16, 2008, by representatives of Universidad de Chile, PSR/COPPE and Universidad Catolica del Peru.

74. "Switching cost" is the minimal price of carbon that would make it financially viable to undertake an investment in a low-emitting technology instead of using a technology that has lower up-front costs, but emits more carbon.

75. World Bank 2008. Environmental Licensing for Hydroelectric Projects in Brazil: A Contribution to the Debate. Brazil Country Management Unit, Report 40995-BR.

76. ESMAP (2007).

77. In South America, Chile and Uruguay are net energy importers, and thus vulnerable to volatility in energy prices

and supplies. However, the dependence on imported hydrocarbons is most acute among Central American and Caribbean countries, including Barbados (86 percent), Dominican Republic (78 percent), Jamaica (86 percent), and Panama (72 percent). ESMAP (2007).

78. ESMAP (2007).

79. See Kojima, M., D. Mitchell, and W. Ward "Considering Trade Policies for Liquid Biofuels," Energy Sector Management Assistance Program Renewable Energy Special Report 004/07, 2007, World Bank.

80. Farrell (2006); Hill and others (2006); Kartha (2006); review of studies reported in Worldwatch Institute (2006) and Kojima, Mitchell, and Ward (2006).

81. Koplow (2006).

82. Mitchell (2008).

83. Farrell (2006); Hill and others (2006); Kartha (2006); review of studies reported in Worldwatch Institute (2006) and Kojima, Mitchell, and Ward (2006).

84. Searchinger and others (2008).

85. Searchinger and others (2008).

86. Zah and others (2007), Gibbs and others (2008).

87. Gibbs and others (2008).

88. Another study that also estimates the carbon payback period concludes that "converting rainforests, peatlands, savannas, or grasslands to produce food-based biofuels in Brazil, Southeast Asia, and the United States creates a 'biofuel carbon debt' by releasing 17 to 420 times more CO_2 than the annual GHG reductions these biofuels provide by displacing fossil fuels." Source: Fargione and others (2008).

89. De Gorter and Tsur (2008).

90. De Gorter and Tsur (2008).

91. The former is 7,225 liters/ha, compared to 3,750 liters/ha. According to Nyberg, J. "SUGAR-BASED ETHANOL International Market Profile." Background paper for the Competitive Commercial Agriculture in Sub–Saharan Africa (CCAA) Study, 2007 FAO and World Bank, citing figures from UNICA. Available at: http://siteresources.worldbank.org/INTAFRICA/Resources/257994-1215457178567/Ethanol_Profile.pdf.

92. De Gorter and Tsur (2008)

93. Smith and others (in press).

94. IPCC (2007).

95. Waste disposal is generally deficient. Only 23 percent of waste collected is disposed in sanitary landfills; another 24 percent goes to controlled landfills, with the remainder ending up in open dumps or courses of water. Pan American Health Organization 2005.

96. West, J. M., and R. V. Salm 2003. "Resistance and Resilience to Coral Bleaching: Implications for Coral Reef Conservation and Management," *Conservation Biology*, 17(Aug), no. 4: 956- 967.

97. Gisselquist, Nash, and Pray (2002) find that overly restrictive seed regulations interfere with technology flow, particularly in some developing countries.

98. P. Michaels, 2008, "Confronting the Political and Scientific Realities of Global Warming," Washington DC: Cato Institute for the Hokkaido G8 Summit.

99. ENSO, a global coupled ocean-atmosphere phenomenon, is associated with floods, droughts, and other disturbances in a range of locations around the world.

100. See, for example, Howitt, R. and E. Pienaar. 2006. "Agricultural Impacts" in J. Smith and R. Mendelsohn (eds.) *The Impact of Climate Change on Regional Systems: A Comprehensive Analysis of California* Edward Elgar Publishing, Northampton, MA. Pp 188–207; Hurd, B., J. Callaway, J. Smith, and P. Kirshen. 1999. "Economics Effects of Climate Change on US Water Resources," in R. Mendelsohn and J. Smith (eds) *The Impact of Climate Change on the United States Economy*. Cambridge University Press, Cambridge, UK, pp. 133–177; Lund, J., T. Zhu, S. Tanaka, M. Jenkins. 2006. "Water Resource Impacts," in J. Smith and R. Mendelsohn (eds.) *The Impact of Climate Change on Regional Systems: A Comprehensive Analysis of California* Edward Elgar Publishing, Northampton, MA. pp 165–187; Strzepek, K., D. Yates, and D. El Quosy. 1996. "Vulnerability assessment of water resources in Egypt to climatic change in the Nile Basin." *Climate Research* 6: 89–95.

101. Mendelsohn, R. 2008, "Impact of Climate Change on the Rio Bravo River." Background paper for this report, July 2.

102. E. Bresnyan and P. Werbrouck, "Value Chains and Small Farmer integration," World Bank, LCSAR, Agriculture for Development series.

103. The CDM that was created under the Kyoto Protocol. This mechanism currently allows industrialized countries to meet some of their climate mitigation commitments by investing in emission reductions in developing countries

104. For example, in one proposal for reducing deforestation rates in the Brazilian Amazon (Nepstad et al. (2007)), financial incentives would be used to partially compensate forest-based local populations and legal private landholders, respectively, for their "forest stewardship" role and forest conservation efforts. In addition, a "government fund" would compensate the government for expenditures above and beyond current outlays, including for the management of public forests, the provision of services to local populations and the monitoring of private forests (including expanded environmental licensing). It is estimated that over a 30-year period, the deforested area could be 490,000 km^2 smaller and avoided emissions 6.3 billion tons of carbon lower than in a business-as-usual scenario estimated by Soares Filho et al. (2006). The overall cost of such a program would be about US$8.2 billion, or about US$1.3 per ton of avoided carbon emissions. It is worth noting, however, that a problem with the proposal of Nepstad et al. (2007) is that it

does not consider it necessary for the financial incentive designed to avoid conversion of forest to soy or cattle ranching to equalize the opportunity cost of the land. The authors cite an ongoing and successful forest protection subsidy program working with local communities and derives the incentive levels from that program.

105. These figures are for the year 2000, the last year for which CAIT (2008) reports emissions of all GHG. Focusing on energy-related CO_2 emissions only yields annual emissions of 0.36 and 0.43 billion tons of CO_2 per year, respectively, for Brazil and Mexico in 2004 (the latest year for which data is available for this type of emissions in CAIT, 2008).

106. Reflecting the country-specific nature of reduction opportunities, of course, other sectors (waste management, agriculture) may be more significant than any of these four in certain countries.

107. FAO (2005).

108. Agrawal, A. 2008. "Livelihoods, Carbon, and Diversity in Community Forests: Tradeoffs or Win-Wins?" Presentation at conference on "Rights, Forests, and Climate Change," October 15–17, 2008, Organized by Rainforest Foundation Norway and the Rights and Resources Foundation. http://rightsandclimate.org/.

109. Chomitz and others (2007).

110. Soarez-Filho and others (2006).

111. The cumulative reduction of particulate matter (PM 2.5) would be of 11,800 tons and that of nitrous oxides of 855,000 tons for the first example, and on the order of 8,000 tons of PM 2.5 and 1,134,000 tons of nitrous oxides for the second. World Bank (2009).

112. Presentations made at CEPAL (Santiago de Chile) on October 16, 2008, by representatives of Fundacion Bariloche, Universidad de Chile, PSR/COPPE, Universidad de los Andes, and Universidad Catolica del Peru.

113. Argentina: The Challenge of Reducing Logistics Costs, 2006; Costa Rica: Country Economic Memorandum: The Challenges for Sustained Growth, 2006; Improving Logistics Costs for Transportation and Trade Facilitation, 2008; Infraestructura Logística y de Calidad para la Competitividad de Colombia, 2006; Brazil: How to Decrease Freight Logistics Costs in Brazil (under preparation).

114. World Bank 2008. Environmental Licensing for Hydroelectric Projects in Brazil: A Contribution to the Debate. Brazil Country Management Unit, Report 40995-BR

115. Rios Roca, A. R., M. Garron B., and P. Cisneros 2005. "Targeting Fuel Subsidies in Latin American and the Caribbean: Analysis and Proposal." Latin American Energy Organization (OLADE), June.

116. Countries are classified as having a relatively high (low) potential when they are above the median LAC country in terms of both (neither) their rate of growth of emissions of a given type and (nor) in terms of the ratio of those emissions to GDP. A *medium* potential is attributed to countries for which the rate of growth of emissions is above the median but the level is not (or vice versa).

117. Definitions of potential are as in table A1 but substituting, in column 1, the levels and rates of growth of the ratio of energy to GDP (over the variables described in table A1); and the level of ratios of emissions to energy instead of that to GDP in the other columns.

118. Definitions of potential are as in table A1.

119. Caribbean community included 15 member countries and 5 associate member countries, totaling 20 countries. Some data are not available for some countries and thus such costs are not estimated in those countries for a specific item. Therefore, the total estimates may be regarded as conservative.